CAMBRIDGE MONOGRAPHS
ON MECHANICS AND APPLIED MATHEMATICS

GENERAL EDITORS

G. K. BATCHELOR, PH.D., F.R.S.
Lecturer in Mathematics in the University of Cambridge

H. BONDI, M.A.
Professor of Applied Mathematics at King's College,
University of London

INTRODUCTION TO FOURIER ANALYSIS AND GENERALISED FUNCTIONS

INTRODUCTION TO
FOURIER ANALYSIS
AND
GENERALISED FUNCTIONS

BY

M. J. LIGHTHILL, F.R.S.

Beyer Professor of Applied Mathematics in the
University of Manchester

CAMBRIDGE
AT THE UNIVERSITY PRESS
1958

PUBLISHED BY
THE SYNDICS OF THE CAMBRIDGE UNIVERSITY PRESS

Bentley House, 200 Euston Road, London, N.W. 1
American Branch: 32 East 57th Street, New York 22, N.Y.

©

CAMBRIDGE UNIVERSITY PRESS
1958

Printed in Great Britain at the University Press, Cambridge
(Brooke Crutchley, University Printer)

CONTENTS

1 Introduction

2 The theory of generalised functions and their Fourier transforms

3 Definitions, properties and Fourier transforms of particular generalised functions

4 The asymptotic estimation of Fourier transforms

5 Fourier series

CHAPTER I

INTRODUCTION

I.I. Scope and purpose of the book

This book grew out of an undergraduate course in the University of Manchester, in which the author attempted to expound the most useful facts about Fourier series and integrals. It seemed to him on planning the course that a satisfactory account must make use of functions like the delta function of Dirac which are outside the usual scope of function theory. Now, Laurent Schwartz in his *Théorie des distributions** has evolved a rigorous theory of these, while Professor Temple has given a version of the theory (*Generalised functions*)† which appears to be more readily intelligible to students. With some slight further simplifications the author found that the theory of generalised functions was accessible to undergraduates in their final year, and that it greatly curtails the labour of understanding Fourier transforms, as well as making available a technique for their asymptotic estimation which seems superior to previous techniques. This is an approach in which the theory of Fourier series appears as a special case, the Fourier transform of a periodic function being a 'row of delta functions'.

The book which grew out of the course therefore covers not only the principal results concerning Fourier transforms and Fourier series, but also serves as an introduction to the theory of generalised functions, whose general properties as well as those useful in Fourier analysis are derived, simply but without any departure from rigorous standards of mathematical proof.

On the other hand, the application of Fourier transforms, or of generalised functions, to the solution of differential, integral or other functional equations is not explicitly treated, nor is the extension to Fourier transforms or generalised functions of more than one variable.

---

* Volumes 1 and 2 (1950–1). Paris: Hermann et Cie.
† *Proc. Roy. Soc.* A, **228**, 175–90 (1955).

1.2. Knowledge assumed of the reader

The book is written for mathematical readers with some interest in, and general knowledge of, methods of mathematical proof (particularly of results concerned with limiting processes). However, little detailed knowledge of particular topics is assumed. The Lebesgue theory of integration is not required; the expression 'absolutely integrable' is frequently used below (where in the Lebesgue theory one would write 'integrable') to remind the reader familiar only with a simpler approach to integration that the integral of the modulus is assumed finite. The theorems that the modulus of an integral is less than the integral of the modulus, and that orders of integration and/or summation can be interchanged in an expression in which convergence is retained when each term is replaced by its modulus, are constantly used, however; and one or two other basic integration theorems are quoted occasionally. The O and o notations are assumed known and understood.*

The functions dealt with are functions of a real variable only (though they have complex *values*). On the other hand, some knowledge of functions of a complex variable is desirable, as use is made on a few occasions of the evaluation of simple definite integrals by contour integration. Familiarity with a few other standard integrals, like the error integral, is also assumed.

All this material is in the ordinary undergraduate course of analysis in British universities. Accordingly, the book should be intelligible both to undergraduates and to working mathematicians whose study of analysis may not have gone much beyond the undergraduate level.

It is, perhaps, unlikely that anyone will come to the book without any previous knowledge of Fourier series or integrals. However, the account of these topics in chapters 2 to 5 is completely self-contained, and in addition some indication is given, in the following sections of this chapter, of some of the principal uses to which Fourier analysis may be put.

---

* Those remembering these notations imperfectly are recommended to read O as 'a term of order at most...', and o as 'a term of order less than...'. This will remind them that '$f = O(g)$', meaning that $|f| < A|g|$ for *some* positive A as the limit is approached, includes the possibility that '$f = o(g)$', meaning that $|f| < \epsilon|g|$ (sufficiently near to the limit) for *any* positive ϵ (however small).

1.3. Fourier series: introductory remarks

A Fourier series is a representation of a periodic function $f(x)$ (say, of period $2l$)* as a linear combination of all those cosine and sine functions which have the same period, say as

$$f(x) = \tfrac{1}{2}a_0 + \sum_{n=1}^{\infty} a_n \cos \frac{n\pi x}{l} + \sum_{n=1}^{\infty} b_n \sin \frac{n\pi x}{l}. \tag{1}$$

Fourier series in this sense are used for analysing oscillations periodic in time, or waveforms periodic in space, and also for representing functions of plane or cylindrical polar co-ordinates, when x in (1) becomes the polar angle θ, and the period $2l$ becomes 2π.

The series can be written, more compactly, in the complex form

$$f(x) = \sum_{n=-\infty}^{\infty} c_n \, \mathrm{e}^{\mathrm{i} n\pi x/l}, \tag{2}$$

where $c_n = \tfrac{1}{2}(a_n - \mathrm{i}b_n)$ for all n, if a_{-n} signifies a_n and b_{-n} signifies $-b_n$ (so that $b_0 = 0$).

One great advantage of expressing a function in terms of cosines and sines, or (even more) in terms of exponentials, is the simple behaviour of these functions under the various operations of analysis, notably differentiation. For example, if a linear partial differential equation has coefficients independent of x, and solutions periodic in x with period $2l$ are required, the series (2) may be used as a solution, and the c_n determined by solving differential equations in which derivatives with respect to x no longer occur.

For example, solutions of Laplace's equation in plane polar co-ordinates r, θ, which are periodic in θ with period 2π (thus, representing solutions which are one-valued in an annulus with centre the origin), may be written as $\sum_{n=-\infty}^{\infty} c_n(r) \, \mathrm{e}^{\mathrm{i} n\theta}$. If we substitute this in

$$\frac{\partial^2 f}{\partial r^2} + \frac{1}{r} \frac{\partial f}{\partial r} + \frac{1}{r^2} \frac{\partial^2 f}{\partial \theta^2} = 0, \tag{3}$$

and assume that we can differentiate term by term, we obtain

$$\sum_{n=-\infty}^{\infty} \left(\frac{\mathrm{d}^2 c_n}{\mathrm{d}r^2} + \frac{1}{r} \frac{\mathrm{d}c_n}{\mathrm{d}r} - \frac{n^2}{r^2} c_n \right) \mathrm{e}^{\mathrm{i} n\theta} = 0. \tag{4}$$

* This means that $f(x) = f(x+2l)$ for all x.

If we assume that expressions of functions by such trigonometrical series are *unique*, then a series which vanishes identically must have vanishing coefficients, which gives a differential equation for c_n, whose general solution is

$$c_n = A_n r^n + B_n r^{-n}. \tag{5}$$

If boundary conditions are available on circles $r = $ constant, for example $f = g(\theta)$ on $r = a$, and say $\partial f / \partial r = h(\theta)$ on $r = b$, we may use these conditions to determine A_n and B_n, provided that $g(\theta)$ and $h(\theta)$ can be represented as Fourier series, say

$$g(\theta) = \sum_{n=-\infty}^{\infty} g_n e^{in\theta}, \quad h(\theta) = \sum_{n=-\infty}^{\infty} h_n e^{in\theta}. \tag{6}$$

Then $A_n a^n + B_n a^{-n} = g_n, \quad nA_n b^{n-1} - nB_n b^{-n-1} = h_n, \tag{7}$

whence A_n and B_n may be determined.

This example is so simple that it could be treated in many different ways (and, in fact, the conclusions just obtained agree with those given by Laurent's theorem), but, clearly, the procedure is directly applicable in more complicated problems, provided always that the boundary conditions from which the c_n are to be determined are given on boundaries where the argument of the Fourier series (here θ) varies independently of the other variables.

The example makes it clear that a satisfactory Fourier-series theory will be one in which term-by-term differentiation and unique determination of the coefficients for a given function are both possible. These two requirements had never been simultaneously satisfied by any of the Fourier-series theories until the 'generalised function' approach given in chapter 5 was developed.

There is a different, and perhaps even commoner, way in which Fourier series are used, namely to represent a function which is not periodic, but instead is defined in the first place only in a restricted interval. The period of the Fourier series is usually taken as twice the length of the interval, and then the series are called 'half-range series'; also, 'quarter-range series' are sometimes used. It will be seen that these series form a restricted class of Fourier series, in which only a proportion of the terms in the 'full-range series' (1) is used.

For example, if a partial differential equation is to be solved in a region part of whose boundary consists of the lines (or planes)

$x=0$ and $x=l$, then the argument is usually presented as follows. The only cosines or sines which satisfy the boundary condition $f=0$ both at $x=0$ and at $x=l$ are $\sin n\pi x/l$ (for $n=1,2,3,\ldots$), so that the half-range series (or 'Fourier sine series')

$$f(x) = \sum_{n=1}^{\infty} b_n \sin \frac{n\pi x}{l} \quad (0<x<l) \tag{8}$$

is appropriate when these are the boundary conditions. Alternatively, if $\partial f/\partial x = 0$ at $x=0$ and $x=l$, the half-range series (or 'Fourier cosine series')

$$f(x) = \tfrac{1}{2}a_0 + \sum_{n=1}^{\infty} a_n \cos \frac{n\pi x}{l} \quad (0<x<l) \tag{9}$$

is appropriate. (In the complex form (2), these cases have c_n pure-imaginary and real, respectively.) Again, if $f=0$ at $x=0$ and $\partial f/\partial x = 0$ at $x=l$, then the quarter-range series

$$f(x) = \sum_{n=1}^{\infty} b_n \sin \frac{(2n-1)\pi x}{2l} \quad (0<x<l), \tag{10}$$

containing an even smaller selection of the terms in a Fourier series of larger period $4l$, is appropriate.

These series can be approached in a slightly different but more useful manner as follows. To satisfy boundary conditions $f(0)=0$ and $f(l)=0$, an *odd periodic* function $f(x)$ of period $2l$ (that is, a function satisfying $f(-x) = -f(x)$ and $f(x+2l) = f(x)$, which by the substitutions $x=0$ and $x=-l$ imply the stated boundary conditions) is introduced. Its Fourier series (1) then contains only odd terms and reduces to (8). Similarly, the boundary conditions $\partial f/\partial x = 0$ at $x=0$ and $x=l$ can be satisfied by using an even periodic function. Finally, the boundary conditions $f=0$ at $x=0$ and $\partial f/\partial x = 0$ at $x=l$ can be satisfied by using an odd periodic function, of period $4l$, which is also an even function of $x-l$; note that only those sine terms of period $4l$ which satisfy the latter condition appear in (10). In each case, naturally, it is the value of the periodic function in the range $0<x<l$ that represents the solution to the problem.

A simple example is now given to show the advantage of replacing boundary conditions wherever possible by conditions of periodicity and parity, even in a case where Fourier series are not used. If a string is stretched between two fixed points $x=0$ and $x=l$ and plucked (that is, released from rest in a given distorted shape), we can imagine an infinite stretched string whose transverse displacement $y = f(x)$ is periodic with period $2l$ and odd and agrees with the given shape for $0 < x < l$ (see fig. 1). If this infinite string were released from rest in this position, the displacement y would remain odd and periodic, and so continue to satisfy the boundary condition $y = 0$ at $x = 0$ and $x = l$. Hence the simple solution of the initial-value problem for the infinite string, namely

$$y = \tfrac{1}{2}\{f(x+ct) + f(x-ct)\}, \tag{11}$$

can be used to give the solution, avoiding the need to consider multiple reflexions from the ends. Similar advantages accrue if other problems of this kind are treated in this way.

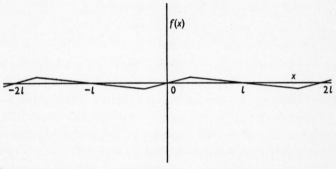

Fig. 1. Illustrating the construction of an odd periodic function $f(x)$ taking given values in $0 < x < l$.

Accordingly, it is best to consider every Fourier series as the Fourier series of a *periodic* function, even in cases where one is interested in the first place only in values over a half- or quarter-period. This is especially advisable in constructing any general theory of Fourier series, since the sum of such a series, if it exists, is certainly periodic!

We may conclude this section by listing the principal aims of a theory of Fourier series. First, one must obtain conditions under which a trigonometrical series like (1) or (2) converges (in the sense which one is using). For example, a sufficient condition for absolute uniform convergence is that $c_n = O(|n|^{-1-\epsilon})$ as $|n| \to \infty$ for some $\epsilon > 0$. In §5.1, however, it will be shown that a necessary and sufficient condition for convergence in the sense of generalised-function theory is that $c_n = O(|n|^N)$ as $|n| \to \infty$ for some N.

Next, if an equation such as (2) is given, one may ask how the c_n may be expressed in terms of the function $f(x)$. For example, when the series is absolutely and uniformly convergent, one may multiply both sides by $e^{-im\pi x/l}$ and integrate term by term from $-l$ to l, giving

$$c_m = \frac{1}{2l} \int_{-l}^{l} f(x) e^{-im\pi x/l} \, dx, \qquad (12)$$

since all the terms of the series vanish except that for which $n = m$. The corresponding problem in the theory of generalised functions is solved in § 5.2.

Note that equation (12) implies

$$a_m = \frac{1}{l} \int_{-l}^{l} f(x) \cos \frac{m\pi x}{l} \, dx, \quad b_m = \frac{1}{l} \int_{-l}^{l} f(x) \sin \frac{m\pi x}{l} \, dx. \qquad (13)$$

Note also that if $f(x)$ is even then

$$a_m = \frac{2}{l} \int_{0}^{l} f(x) \cos \frac{m\pi x}{l} \, dx, \quad b_m = 0, \qquad (14)$$

which are the equations for a Fourier cosine series, while if $f(x)$ is odd then

$$a_m = 0, \quad b_m = \frac{2}{l} \int_{0}^{l} f(x) \sin \frac{m\pi x}{l} \, dx, \qquad (15)$$

which are the equations for a Fourier sine series.

Next, one may choose an arbitrary periodic function $f(x)$, and ask under what conditions equation (2) will hold with the c_n given by (12), or by the equivalent expression in the theory one is using. Elaborate tests must be satisfied in the ordinary convergence theory for this to be so, but it will be proved in § 5.3 that the equation holds for all periodic generalised functions.

Finally, one may ask the questions already mentioned concerning term-by-term differentiability and uniqueness. These are the properties which are most notably lacking in the 'convergence' and 'summability' theories, respectively;* but both results are almost trivial in the generalised-function theory.

* For these classical theories see, for example, G. H. Hardy and W. W. Rogosinski, *Fourier Series* (2nd edition, 1950), Cambridge University Press.

1.4. Fourier transforms: introductory remarks

The Fourier integral may be regarded as the formal limit of the Fourier series as the period tends to infinity. Thus, if $f(x)$ is any function of x in the whole range $(-\infty, \infty)$, one can form a periodic function $f_l(x)$ of period $2l$ which agrees with $f(x)$ in the range $(-l, l)$. The Fourier series (2) of $f_l(x)$, with the expressions (12) for the c_n, can be written in the form

$$f_l(x) = \sum_{n=-\infty}^{\infty} e^{in\pi x/l} g_l\left(\frac{n}{2l}\right)\frac{1}{2l}, \quad \text{where} \quad g_l(y) = \int_{-l}^{l} e^{-2\pi i x y} f(x)\,dx.$$

(16)

A formal limit as $l \to \infty$, where in the series $n/2l$ is written y, and the difference between successive values of y is written dy, is

$$f(x) = \int_{-\infty}^{\infty} e^{2\pi i x y} g(y)\,dy, \quad \text{where} \quad g(y) = \int_{-\infty}^{\infty} e^{-2\pi i x y} f(x)\,dx,$$

(17)

since in the limit the periodic function $f_l(x)$ becomes $f(x)$ everywhere.

Under these circumstances the function $g(y)$ is often called the Fourier transform (F.T.) of $f(x)$. Then the first of equations (17) may be regarded as stating that $f(y)$ is the F.T. of $g(-x)$.

The reader should be warned, however, that no general agreement has been reached on where the 2π's in the definition of Fourier transforms should be put. They can be taken out of the exponentials in equations (17), and a $1/2\pi$ factor inserted before the first integral but not the second (or vice versa), or else a factor $1/\sqrt{(2\pi)}$ can in this case be inserted before each (to maintain as much symmetry as possible). All the different notations which have been used have some advantages. Here we follow Temple, and many modern authors, in including the 2π in the exponent, so that the exponential multiplying $g(y)$ represents an oscillation with y as frequency (or wave number, according as x represents time or space), rather than as 'radian frequency' or 'radian wave number'. However, the results in this book are easily changed into one of the other notations by a slight change of variable.

A vast literature* has been devoted to the determination of con-

---

* See, for example, E. C. Titchmarsh, *Introduction to the theory of Fourier integrals* (1937), Oxford University Press.

ditions on $f(x)$ sufficient for equations (17) to be true with a given interpretation of the integrals. Even for a fixed interpretation, many alternative sets of sufficient conditions are necessary if one wishes to apply the equations at all widely, because relaxation of one condition to admit some desired function requires usually the strengthening of some other condition, which in turn excludes some other function. All these difficulties disappear when generalised functions are used, since every generalised function $f(x)$ has a Fourier transform $g(y)$ which is also a generalised function, and the F.T. of $g(-x)$ is $f(y)$. In the latter theory it is found convenient to proceed in an order different from that adopted in this chapter, and to treat the properties of Fourier series as a special case of the properties of Fourier transforms.

The Fourier integral is used to analyse non-periodic functions of x in the range $(-\infty, \infty)$, as linear combinations of exponential functions. Such an analysis is useful for much the same reasons as with Fourier series. For example, it is effective in treating linear partial differential equations with coefficients independent of x, subject to boundary conditions given on boundaries where x varies from $-\infty$ to ∞ independently of the other variables. To apply these boundary conditions, it is necessary to be able to express any functions occurring in them as Fourier integrals. Here, a difficulty used to be that quite simple functions (for example, a constant!) have no Fourier transform in ordinary function theory. This difficulty disappears in the theory of generalised functions, in which, for example, the F.T. of 1 is the delta function of Dirac.

'Half-range' Fourier integrals are also used, along much the same lines as with Fourier series. Thus, if $f(x)$ is to be determined in the range $(0, \infty)$ subject to a condition $f(0)=0$, one may seek an odd function $f(x)$ in the full range $(-\infty, \infty)$, which coincides with $f(x)$ in $(0, \infty)$. Its F.T. $g(y)$, by (17), may be written as

$$g(y) = -2i \int_0^\infty f(x) \sin 2\pi xy \, dx, \qquad (18)$$

which is also an odd function, so that the expression for $f(x)$ in terms of $g(y)$ becomes

$$f(x) = 2i \int_0^\infty g(y) \sin 2\pi xy \, dy. \qquad (19)$$

The integrals in (18) and (19) are called Fourier sine integrals.

Similarly, if $f(x)$ is to be determined in $(0, \infty)$ subject to a condition $\partial f/\partial x = 0$ when $x = 0$, one may seek an even function $f(x)$ in the full range $(-\infty, \infty)$ which coincides with $f(x)$ in $(0, \infty)$. Its F.T. may be written as

$$g(y) = 2 \int_0^\infty f(x) \cos 2\pi xy \, dx, \tag{20}$$

which is also an even function, and the expression for $f(x)$ in terms of $g(y)$ becomes

$$f(x) = 2 \int_0^\infty g(y) \cos 2\pi xy \, dy. \tag{21}$$

The integrals in (20) and (21) are called Fourier cosine integrals.

As with Fourier series, no special theory is needed for Fourier sine integrals and Fourier cosine integrals. They should be regarded simply as what is obtained by taking Fourier transforms of odd and even functions respectively.

In many cases, especially when it is not possible to evaluate a Fourier transform explicitly in terms of tabulated functions, it is useful to have a technique for evaluating the asymptotic behaviour of the F.T. $g(y)$ as $|y| \to \infty$, in terms of the behaviour of $f(x)$ near its singularities. It is difficult to find a comprehensive account of this technique in the literature, and since the theory becomes particularly simple when generalised functions are used, a substantial fraction of this book has been devoted to expounding it. This theory can also be applied without change (§ 5.5) to the problem of determining the asymptotic behaviour as $|n| \to \infty$ of the coefficients c_n in the Fourier series for a given function.

1.5. Generalised functions: introductory remarks

The first 'generalised function' to be introduced was Dirac's 'delta function' $\delta(x)$, which has the property

$$\int_{-\infty}^\infty \delta(x) F(x) \, dx = F(0) \tag{22}$$

for any suitably continuous function $F(x)$. No function in the ordinary sense has the property (22), but one can imagine a sequence of functions (see, for example, fig. 2) which have progressively taller and thinner peaks at $x = 0$, with the area under the curve remaining equal to 1, while the value of the function tends to 0 at

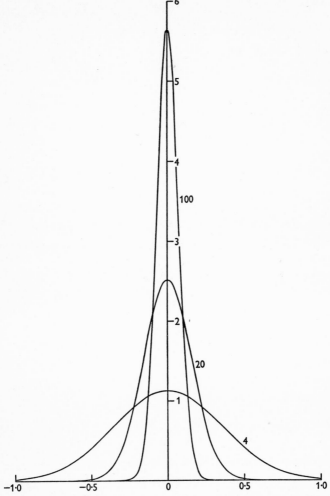

Fig. 2. Functions in the sequence (chapter 2, example 6) used to define $\delta(x)$; the number n (for $n=4$, 20, 100) is attached to the graph of the nth function.

every point, except $x=0$ where it tends to infinity. In the limit, this sequence would have the property (22).

Again, one might 'differentiate' $\delta(x)$ to obtain a function $\delta'(x)$ with the property

$$\int_{-\infty}^{\infty} \delta'(x)\,F(x)\,dx = -\int_{-\infty}^{\infty} \delta(x)\,F'(x)\,dx = -F'(0) \qquad (23)$$

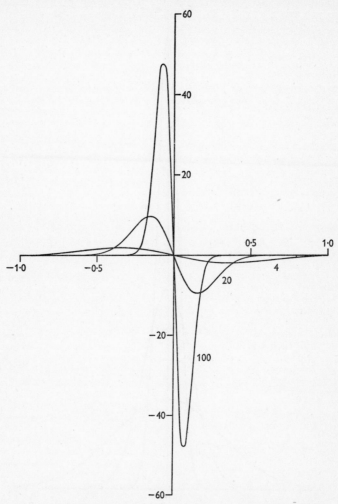

Fig. 3. Functions in the sequence used to define $\delta'(x)$, being the derivatives of those graphed in fig. 2.

for any continuously differentiable function $F(x)$. Behaviour like (23) can again be realised in the limit of a sequence of functions (for example, the derivatives of those in the sequence used to represent $\delta(x)$; these are graphed in fig. 3).

Physically, $\delta(x)$ can be regarded as that distribution of charge along the x-axis which one speaks of as a unit point charge at the origin. Similarly, $\delta'(x)$ corresponds to a dipole of unit electric moment, since as a special case of (23) we have

$$\int_{-\infty}^{\infty} x\delta'(x)\,dx = -1. \tag{24}$$

Thus, these generalised functions correspond to familiar physical idealisations.

The definition by means of a sequence is in fact that to be adopted in chapter 2, following Temple (who in this point follows Mikusínski). Alternative definitions are not considered here; the reader is referred for a historical account of the subject to Temple, *J. Lond. Math. Soc.* **28**, 134–48 (1953).

In defining generalised functions by means of sequences, one must define under what circumstances two sequences constitute the same generalised function. For this purpose one multiplies each member of a sequence by a 'test function' $F(x)$, as in (22) or (23), integrates from $-\infty$ to ∞, and takes the limit. If the same result emerges for each sequence whatever 'test function' is used, the sequences are said to define the same generalised function.

Different classes of test functions, and of functions admitted for membership of the sequences, can be used in different versions of the theory, but there is only one satisfactory choice when Fourier transforms are to be used. We will not tolerate any restrictions on the differentiability of generalised functions; therefore, as an equation like (23) indicates, both the functions admitted as members of sequences, and the test functions, must have derivatives of all orders. But the classes must be such that the Fourier transform of a member of either class is also a member of that class. The widest class satisfying both these conditions is that introduced in definition 1 (at the beginning of the next chapter). Therefore, we allow both the members of the sequences and the test functions to be arbitrary members of this class. We call members of this class 'good functions', as a graphic term, which it seems desirable to introduce in preference to 'test functions' since it is not only the test functions which are to be restricted to membership of this class.

The development of the theory of generalised functions in chapter 2 follows closely the lines suggested by Temple, but with the omission of a clause in the definition of a generalised function which requires the integral of its product with a test function to be a 'continuous linear functional' of the test function in a certain sense. To keep this restriction would introduce considerable complications into the proofs in the earlier stages of the theory, and it appears possible to proceed satisfactorily without it. Only at one point (theorem 24 in chapter 5) has it been necessary to prove a result, which is actually a special case of Schwartz's general theorem that a linear functional *must* (in fact) be continuous in this sense, and the short hold-up here seems amply compensated for by the increased simplicity of the general theory. Another small difference from Temple's papers is that the author has not found the concept of the indefinite integral of a generalised function valuable, and has accordingly omitted it.

Although the delta function was the first generalised function to be introduced, the methods of attaching values to integrals and series which are introduced in the theory had much earlier forerunners, like Cauchy's 'principal value' and Hadamard's 'finite part' of an improper integral, and the theories of 'summability' of series. The connexions with these topics are briefly mentioned in chapters 3 and 5.

CHAPTER 2

THE THEORY OF GENERALISED FUNCTIONS AND THEIR FOURIER TRANSFORMS

2.1. Good functions and fairly good functions

DEFINITION 1. *A good function is one which is everywhere differentiable any number of times and such that it and all its derivatives are $O(|x|^{-N})$ as $|x| \to \infty$ for all N.*

EXAMPLE 1. e^{-x^2} is a good function.

DEFINITION 2. *A fairly good function is one which is everywhere differentiable any number of times and such that it and all its derivatives are $O(|x|^N)$ as $|x| \to \infty$ for some N.*

EXAMPLE 2. Any polynomial is a fairly good function.

THEOREM 1. *The derivative of a good function is a good function. The sum of two good functions is a good function. The product of a fairly good function and a good function is a good function.*

The proof is left to the reader.

THEOREM 2. *If $f(x)$ is a good function, then the Fourier transform (F.T.) of $f(x)$, namely*

$$g(y) = \int_{-\infty}^{\infty} f(x)\, e^{-2\pi i x y}\, dx, \tag{1}$$

is a good function.

PROOF. Differentiation p times and integration by parts N times shows that

$$\left| g^{(p)}(y) \right| = \left| \frac{1}{(2\pi i y)^N} \int_{-\infty}^{\infty} \frac{d^N}{dx^N}\{(-2\pi i x)^p f(x)\}\, e^{-2\pi i x y}\, dx \right|$$

$$\leqslant \frac{(2\pi)^{p-N}}{|y|^N} \int_{-\infty}^{\infty} \left| \frac{d^N}{dx^N}\{x^p f(x)\} \right| dx = O(|y|^{-N}), \tag{2}$$

which proves the theorem.

THEOREM 3. *If $f(x)$ is a good function with F.T. $g(y)$, then the F.T. of $f'(x)$ is $2\pi i y g(y)$, and the F.T. of $f(ax+b)$ is $|a|^{-1} e^{2\pi i b y/a} g(y/a)$.*

The proof is left to the reader, who should note the special cases $a = 1$, $a = -1$ and $b = 0$.

THEOREM 4 (Fourier's inversion theorem for good functions). *If $g(y)$ is the F.T. of a good function $f(x)$, then $f(y)$ is the F.T. of $g(-x)$.*

PROOF. Any of the standard proofs of Fourier's inversion theorem applies without any difficulty to good functions. A simple version is to prove by elementary manipulation that the F.T. of $e^{-\epsilon x^2} g(-x)$ differs from $f(y)$ by

$$\left| \int_{-\infty}^{\infty} \left(\frac{\pi}{\epsilon} \right)^{\frac{1}{2}} e^{-\pi^2 (y - t^2)/\epsilon} \{ f(t) - f(y) \} \, dt \right|$$
$$\leqslant \max |f'(x)| \int_{-\infty}^{\infty} \left(\frac{\pi}{\epsilon} \right)^{\frac{1}{2}} e^{-\pi^2 (y - t)^2/\epsilon} |y - t| \, dt = O(\epsilon^{\frac{1}{2}}), \quad (3)$$

and then let $\epsilon \to 0$.

THEOREM 5 (Parseval's theorem for good functions). *If $f_1(x)$ and $f_2(x)$ are good functions, and $g_1(y)$ and $g_2(y)$ are their F.T.'s, then*

$$\int_{-\infty}^{\infty} g_1(y) g_2(y) \, dy = \int_{-\infty}^{\infty} f_1(-x) f_2(x) \, dx. \quad (4)$$

PROOF. Both sides can be written as the absolutely convergent double integral

$$\int_{-\infty}^{\infty} \int_{-\infty}^{\infty} g_1(y) f_2(x) e^{-2\pi i x y} \, dx \, dy, \quad (5)$$

by theorem 4.

NOTE. This theorem will also be used (in theorem 11 below) in a case when $f_1(x)$ is a good function and $f_2(x)$ is any function absolutely integrable from $-\infty$ to ∞. The proof stands word for word in this case, since the double integral remains absolutely convergent.

2.2. Generalised functions. The delta function and its derivatives

DEFINITION 3. *A sequence $f_n(x)$ of good functions is called regular if, for any good function $F(x)$ whatever, the limit*

$$\lim_{n \to \infty} \int_{-\infty}^{\infty} f_n(x) F(x) \, dx \quad (6)$$

exists.

EXAMPLE 3. The sequence $f_n(x) = e^{-x^2/n^2}$ is regular. $\left(\text{The limit in this case is } \int_{-\infty}^{\infty} F(x) \, dx. \right)$

DEFINITION 4. *Two regular sequences of good functions are called equivalent if, for any good function $F(x)$ whatever, the limit (6) is the same for each sequence.*

EXAMPLE 4. The sequence e^{-x^4/n^4} is equivalent to the sequence e^{-x^2/n^2}.

DEFINITION 5. *A generalised function $f(x)$ is defined as a regular sequence $f_n(x)$ of good functions, but two generalised functions are said to be equal if the corresponding regular sequences are equivalent. Thus, each generalised function is really the class of all regular sequences equivalent to a given regular sequence. The integral*

$$\int_{-\infty}^{\infty} f(x)\,F(x)\,\mathrm{d}x \tag{7}$$

of the product of a generalised function $f(x)$ and a good function $F(x)$ is defined as

$$\lim_{n \to \infty} \int_{-\infty}^{\infty} f_n(x)\,F(x)\,\mathrm{d}x. \tag{8}$$

This is permissible because the limit is the same for all equivalent sequences $f_n(x)$.

EXAMPLE 5. The sequence e^{-x^2/n^2} and all equivalent sequences define a generalised function $I(x)$ such that

$$\int_{-\infty}^{\infty} I(x)\,F(x)\,\mathrm{d}x = \int_{-\infty}^{\infty} F(x)\,\mathrm{d}x. \tag{9}$$

This generalised function $I(x)$ will be denoted more simply by 1.

EXAMPLE 6. The sequences equivalent to $e^{-nx^2}(n/\pi)^{\frac{1}{2}}$ define a generalised function $\delta(x)$ such that

$$\int_{-\infty}^{\infty} \delta(x)\,F(x)\,\mathrm{d}x = F(0). \tag{10}$$

PROOF. If $F(x)$ is any good function,

$$\left| \int_{-\infty}^{\infty} e^{-nx^2}(n/\pi)^{\frac{1}{2}}\,F(x)\,\mathrm{d}x - F(0) \right|$$

$$= \left| \int_{-\infty}^{\infty} e^{-nx^2}(n/\pi)^{\frac{1}{2}}\{F(x) - F(0)\}\,\mathrm{d}x \right|$$

$$\leqslant \max|F'(x)| \int_{-\infty}^{\infty} e^{-nx^2}(n/\pi)^{\frac{1}{2}}\,|x|\,\mathrm{d}x$$

$$= (\pi n)^{-\frac{1}{2}}\max|F'(x)| \to 0 \quad \text{as} \quad n \to \infty.$$

DEFINITION 6. *If two generalised functions $f(x)$ and $h(x)$ are defined by sequences $f_n(x)$ and $h_n(x)$, then their sum $f(x) + h(x)$ is defined by the sequence $f_n(x) + h_n(x)$. Also, the derivative $f'(x)$ is defined by the sequence $f'_n(x)$. Also, $f(ax + b)$ is defined by the sequence $f_n(ax + b)$. Also, $\phi(x)f(x)$, where $\phi(x)$ is a fairly good function, is defined by the sequence $\phi(x)f_n(x)$. Also, the F.T. $g(y)$ of $f(x)$ is defined by the sequence $g_n(y)$, where $g_n(y)$ is the F.T. of $f_n(x)$.*

PROOF OF CONSISTENCY. In each item of this definition we must verify (i) that the sequence named is a sequence of good functions, but this follows at once from theorems 1 and 2; (ii) that the sequence named is a regular sequence; and (iii) that different choices of equivalent regular sequences to define the generalised functions f and h lead to equivalent sequences defining the new generalised function. Now, as regards the first item, for any good function $F(x)$

$$\lim_{n \to \infty} \int_{-\infty}^{\infty} \{f_n(x) + h_n(x)\} F(x) \, dx$$
$$= \lim_{n \to \infty} \int_{-\infty}^{\infty} f_n(x) F(x) \, dx + \lim_{n \to \infty} \int_{-\infty}^{\infty} h_n(x) F(x) \, dx, \quad (11)$$

and so the limit on the left exists, verifying (ii). Also, the limits on the right are independent of which of the different equivalent sequences of good functions f_n and h_n are used to define f and h. Hence all resulting sequences $f_n + h_n$ are equivalent, verifying (iii). Again,

$$\lim_{n \to \infty} \int_{-\infty}^{\infty} f'_n(x) F(x) \, dx = - \lim_{n \to \infty} \int_{-\infty}^{\infty} f_n(x) F'(x) \, dx, \quad (12)$$

and, since $F'(x)$ is a good function (by theorem 1), the limit on the right exists and is the same (by definitions 3 and 4) for all equivalent regular sequences $f_n(x)$. Hence all the sequences $f'_n(x)$ are equivalent and regular, as was to be proved. Precisely the same argument applies to

$$\lim_{n \to \infty} \int_{-\infty}^{\infty} f_n(ax + b) F(x) \, dx = \frac{1}{|a|} \lim_{n \to \infty} \int_{-\infty}^{\infty} f_n(x) F\left(\frac{x - b}{a}\right) dx, \quad (13)$$

$$\lim_{n \to \infty} \int_{-\infty}^{\infty} \{\phi(x) f_n(x)\} F(x) \, dx = \lim_{n \to \infty} \int_{-\infty}^{\infty} f_n(x) \{\phi(x) F(x)\} \, dx \quad (14)$$

and $$\lim_{n \to \infty} \int_{-\infty}^{\infty} g_n(y) G(y) \, dy = \lim_{n \to \infty} \int_{-\infty}^{\infty} f_n(x) F(-x) \, dx, \quad (15)$$

where in (15) $G(y)$ is the F.T. of $F(x)$ and theorem 5 has been used. This completes the proof that addition, differentiation, linear substitution, multiplication by a fairly good function and Fourier transformation can each be applied to any generalised function, and that the result in each case is still a generalised function.*

EXAMPLE 7. The F.T. of $\delta(x)$ is 1.

PROOF. The F.T. of $e^{-nx^2}(n/\pi)^{\frac{1}{2}}$ is easily found to be $e^{-\pi^2 y^2/n}$, which is obviously one of the sequences defining the generalised function 1.

THEOREM 6. *Under the conditions of definition 6, we have for any good function* $F(x)$ *(with F.T.* $G(y)$*)*

$$\left.\begin{array}{l}\displaystyle\int_{-\infty}^{\infty} f'(x)\,F(x)\,\mathrm{d}x = -\int_{-\infty}^{\infty} f(x)\,F'(x)\,\mathrm{d}x, \\[2mm] \displaystyle\int_{-\infty}^{\infty} f(ax+b)\,F(x)\,\mathrm{d}x = \frac{1}{|a|}\int_{-\infty}^{\infty} f(x)\,F\!\left(\frac{x-b}{a}\right)\,\mathrm{d}x, \\[2mm] \displaystyle\int_{-\infty}^{\infty} \{\phi(x)\,f(x)\}\,F(x)\,\mathrm{d}x = \int_{-\infty}^{\infty} f(x)\,\{\phi(x)\,F(x)\}\,\mathrm{d}x, \\[2mm] \displaystyle\int_{-\infty}^{\infty} g(y)\,G(y)\,\mathrm{d}y = \int_{-\infty}^{\infty} f(x)\,F(-x)\,\mathrm{d}x. \end{array}\right\} \quad (16)$$

PROOF. These equations follow at once from equations (12) to (15).

EXAMPLE 8. If $F(x)$ is any good function,

$$\int_{-\infty}^{\infty} \delta^{(n)}(x)\,F(x)\,\mathrm{d}x = (-1)^n\,F^{(n)}(0). \quad (17)$$

PROOF. This follows by n-fold application of the first of equations (16) ('integration by parts'), followed by equation (10).

At this stage one can, if thought necessary, prove a whole corpus of results like

$$\frac{\mathrm{d}}{\mathrm{d}x}\{f(x)+h(x)\} = f'(x)+h'(x),$$

$$\frac{\mathrm{d}}{\mathrm{d}x}\{\phi(x)\,f(x)\} = \phi'(x)f(x)+\phi(x)\,f'(x), \qquad \frac{\mathrm{d}}{\mathrm{d}x}f(ax+b) = af'(ax+b),$$

$$\phi(ax+b)\,f(ax+b) = h(ax+b) \quad \text{if} \quad \phi(x)\,f(x) = h(x),$$

---

* On the other hand, there is no satisfactory definition of the product of two generalised functions; for example, in the notation of definition 6, $f_n(x)\,h_n(x)$ is not in general a regular sequence.

etc., which are hardly worth dignifying with the name of theorem since it is almost impossible to imagine reasonable definitions under which they would not be true; they will be used without reference, and the proof is short and easy in each case. The longest is as follows. If $F(x)$ is any good function, then

$$\int_{-\infty}^{\infty} F(x) \frac{d}{dx} \{\phi(x) f(x)\} \, dx$$

$$= -\int_{-\infty}^{\infty} F'(x) \, \phi(x) f(x) \, dx$$

$$= -\int_{-\infty}^{\infty} \frac{d}{dx} \{F(x) \, \phi(x)\} f(x) \, dx + \int_{-\infty}^{\infty} F(x) \, \phi'(x) f(x) \, dx$$

$$= \int_{-\infty}^{\infty} F(x) \{\phi(x) f'(x) + \phi'(x) f(x)\} \, dx.$$

The more useful results on Fourier transforms are, however, collected into a theorem.

THEOREM 7. *If $f(x)$ is a generalised function with F.T. $g(y)$, then the F.T. of $f(ax+b)$ is $|a|^{-1} e^{2\pi i b y/a} g(y/a)$. Also, the F.T. of $f'(x)$ is $2\pi i y g(y)$. Finally* (Fourier's inversion theorem for generalised functions), *$f(y)$ is the F.T. of $g(-x)$.*

PROOF. This powerful theorem follows at once from definition 6 and theorems 3 and 4. For example, to prove the inversion theorem, let the sequence $f_n(x)$ define $f(x)$; then, by definition 6, $g_n(y)$, the F.T. of $f_n(x)$, defines $g(y)$, whence $g_n(-x)$ defines $g(-x)$. But, by theorem 4, the F.T. of $g_n(-x)$ is $f_n(y)$. Hence the F.T. of $g(-x)$ is $f(y)$.

EXAMPLE 9. The F.T. of $\delta(x-c)$ is $e^{-2\pi i c y}$, by example 7 and theorem 7. Hence, by the last part of theorem 7, the F.T. of $e^{2\pi i c x}$ is $\delta(y-c)$.

THEOREM 8. *If $f(x)$ is a generalised function and $f'(x)=0$, then $f(x)$ is a constant (that is, $f(x)$ is equal to a constant times the generalised function 1).*

PROOF. If $F(x)$ is a good function, then clearly

$$F_1(x) = \int_{-\infty}^{x} F(x) \, dx$$

is a good function if and only if $\int_{-\infty}^{\infty} F(x)\,dx = 0$. It follows that

$$F_2(x) = \int_{-\infty}^{x} \left\{ F(x) - \frac{e^{-x^2}}{\sqrt{\pi}} \int_{-\infty}^{\infty} F(t)\,dt \right\} dx \qquad (18)$$

is always a good function, since the function in curly brackets is a good function whose integral from $-\infty$ to ∞ vanishes. Hence,

$$\int_{-\infty}^{\infty} f(x)\,F(x)\,dx = \left(\int_{-\infty}^{\infty} f(x) \frac{e^{-x^2}}{\sqrt{\pi}}\,dx \right) \int_{-\infty}^{\infty} F(t)\,dt + \int_{-\infty}^{\infty} f(x)\,F_2'(x)\,dx$$

$$= C \int_{-\infty}^{\infty} F(x)\,dx - \int_{-\infty}^{\infty} f'(x)\,F_2(x)\,dx, \qquad (19)$$

where C is a constant. The last integral in (19) vanishes, since $f'(x) = 0$. Hence, $f(x) = C$.

THEOREM 9. *If $g(y)$ is a generalised function and $yg(y) = 0$, then $g(y)$ is a constant times $\delta(y)$.*

PROOF. This theorem follows immediately from theorem 8 by taking Fourier transforms (using the second part of theorem 7 and example 7). It can also be proved independently as follows. If $G(y)$ is a good function, then

$$G_1(y) = \frac{G(y) - G(0)\,e^{-y^2}}{y} \qquad (20)$$

is a good function. Hence

$$\int_{-\infty}^{\infty} g(y)\,G(y)\,dy = G(0) \int_{-\infty}^{\infty} g(y)\,e^{-y^2}\,dy$$

$$+ \int_{-\infty}^{\infty} yg(y)\,G_1(y)\,dy = CG(0), \qquad (21)$$

where C is a constant, and the last integral vanishes because $yg(y) = 0$. Hence $g(y) = C\delta(y)$.

2.3. Ordinary functions as generalised functions

DEFINITION 7. *If $f(x)$ is a function of x in the ordinary sense, such that $(1 + x^2)^{-N} f(x)$ is absolutely integrable from $-\infty$ to ∞ for some N, then the generalised function $f(x)$ is defined by a sequence $f_n(x)$ such that for any good function $F(x)$*

$$\lim_{n \to \infty} \int_{-\infty}^{\infty} f_n(x)\,F(x)\,dx = \int_{-\infty}^{\infty} f(x)\,F(x)\,dx. \qquad (22)$$

NOTE. The integral on the right is the integral in the ordinary sense, which exists as the integral of the product of $(1+x^2)^{-N}f(x)$, which is absolutely integrable, and $(1+x^2)^N F(x)$, which is a good function. When the generalised function $f(x)$ has been defined, this integral has a meaning also in the theory of generalised functions, and equation (22) states that these two meanings are the same.

PROOF OF CONSISTENCY. It must be shown that such a sequence exists. We take*

$$f_n(x) = \int_{-\infty}^{\infty} f(t)\, S\{n(t-x)\}\, n\, e^{-t^2/n^2}\, dt, \qquad (23)$$

where the 'smudge function' $S(y)$ is any good function which is zero for $|y| \geqslant 1$ and positive for $|y| < 1$ and satisfies $\int_{-1}^{1} S(y)\, dy = 1$. For example, $S(y)$ may be taken as

$$e^{-1/(1-y^2)} \left\{ \int_{-1}^{1} e^{-1/(1-z^2)}\, dz \right\}^{-1} \qquad (24)$$

for $|y| < 1$ and zero for $|y| \geqslant 1$; note that on this definition all the derivatives of $S(y)$ exist even at $y = \pm 1$ (they are all zero there).

We must now prove that the $f_n(x)$ are good functions, and that equation (22) is satisfied. First,

$$|f_n^{(p)}(x)| = \left| \int_{-\infty}^{\infty} f(t)\,(-n)^p\, S^{(p)}\{n(t-x)\}\, n\, e^{-t^2/n^2}\, dt \right|$$

$$\leqslant n^{p+1} \max | S^{(p)}(y) |\, e^{-(|x|-1)^2/n^2} \{1 + (|x|+1)^2\}^N$$

$$\times \int_{-\infty}^{\infty} (1+t^2)^{-N} |f(t)|\, dt$$

$$= O(|x|)^{-M} \quad \text{as} \quad |x| \to \infty \text{ for all } M, \qquad (25)$$

where we have used the fact that where the integrand is non-zero

$$|x| - 1 < |t| < |x| + 1.$$

---

* The S factor in the integral 'smudges' f over a small interval $(x-n^{-1}, x+n^{-1})$. The e^{-t^2/n^2} factor makes it 'good' at infinity.

Secondly,

$$\left| \int_{-\infty}^{\infty} f_n(x) F(x) \, dx - \int_{-\infty}^{\infty} f(x) F(x) \, dx \right|$$

$$= \left| \int_{-1}^{1} S(y) \, dy \left\{ \int_{-\infty}^{\infty} f(t) \, e^{-t^2/n^2} F\left(t - \frac{y}{n}\right) dt - \int_{-\infty}^{\infty} f(t) F(t) \, dt \right\} \right|$$

$$\leqslant \max_{|y|<1} \left| \int_{-\infty}^{\infty} f(t) \, e^{-t^2/n^2} \left\{ F\left(t - \frac{y}{n}\right) - F(t) \right\} dt \right.$$

$$\left. - \int_{-\infty}^{\infty} f(t) F(t) \left(1 - e^{-t^2/n^2}\right) dt \right|$$

$$\leqslant \int_{-\infty}^{\infty} |f(t)| \left\{ \frac{1}{n} \max_{|x-t|<1} |F'(x)| \right\} dt + \int_{-\infty}^{\infty} |f(t) F(t)| \frac{1+t^2}{n^2} \, dt$$

$$\leqslant \frac{1}{n} \int_{-\infty}^{\infty} |f(t)| \frac{A}{(1+t^2)^N} \, dt + \frac{1}{n^2} \int_{-\infty}^{\infty} |f(t)| \frac{B}{(1+t^2)^N} \, dt$$

$$\to 0 \quad \text{as} \quad n \to \infty, \tag{26}$$

where A and B are constants and the facts that $F(x)$ is a good function and that $(1+t^2)^{-N} f(t)$ is absolutely integrable have again been used. This completes the proof of consistency.

Definition 7 increases enormously the range of generalised functions available to us. Not only can all ordinary functions $f(x)$ with $(1+x^2)^{-N} f(x)$ absolutely integrable from $-\infty$ to ∞ be used as generalised functions, but one can obtain from them new generalised functions by differentiation in accordance with definition 6.

EXAMPLE 10. The discontinuous function sgn x, which is 1 for $x > 0$ and -1 for $x < 0$, is a generalised function, and

$$d \, \mathrm{sgn} \, x / dx = 2\delta(x).$$

PROOF. sgn x satisfies the condition of definition 7 (with $N = 1$) and, for any good function $F(x)$,

$$\int_{-\infty}^{\infty} \frac{d \, \mathrm{sgn} \, x}{dx} F(x) \, dx = - \int_{-\infty}^{\infty} \mathrm{sgn} \, x F'(x) \, dx$$

$$= - \int_{0}^{\infty} F'(x) \, dx + \int_{-\infty}^{0} F'(x) \, dx = 2F(0). \tag{27}$$

THEOREM 10. *If $f(x)$ is an ordinary differentiable function such that both $f(x)$ and $f'(x)$ satisfy the condition of definition 7, then the derivative of the generalised function formed from $f(x)$ is the generalised function formed from $f'(x)$.*

PROOF. This theorem, which shows that the notation $f'(x)$ can be used without risk of confusion, follows from the fact that both definitions of it satisfy

$$\int_{-\infty}^{\infty} f'(x)\,F(x)\,\mathrm{d}x = -\int_{-\infty}^{\infty} f(x)\,F'(x)\,\mathrm{d}x \qquad (28)$$

for any good function $F(x)$. With the second definition, equation (28) assumes that $f(x)\,F(x) \to 0$ as $x \to +\infty$ or $x \to -\infty$. However, the product must tend to *some* finite limit in each case, since

$$\int_{-\infty}^{\infty} f(x)\,F'(x)\,\mathrm{d}x + \int_{-\infty}^{\infty} f'(x)\,F(x)\,\mathrm{d}x$$

exists, whence both limits must be zero since $\int_{-\infty}^{\infty} f(x)\,F(x)\,\mathrm{d}x$ exists.

THEOREM 11. *If $f(x)$ is an ordinary function which is absolutely integrable from $-\infty$ to ∞, so that its F.T. $g(y)$ in the ordinary sense exists, then the F.T. of the generalised function $f(x)$ is the generalised function $g(y)$.*

PROOF. The generalised function $g(y)$ exists because $g(y)$ satisfies the condition of definition 7 with $N = 1$; the integral

$$\int_{-\infty}^{\infty} \frac{\mathrm{d}y}{1+y^2} \int_{-\infty}^{\infty} f(x)\,\mathrm{e}^{-2\pi \mathrm{i}xy}\,\mathrm{d}x \qquad (29)$$

remains convergent when each term is replaced by its modulus. But, by the note following theorem 5, the ordinary function $g(y)$ satisfies

$$\int_{-\infty}^{\infty} g(y)\,G(y)\,\mathrm{d}y = \int_{-\infty}^{\infty} f(x)\,F(-x)\,\mathrm{d}x \qquad (30)$$

for any good function $F(x)$ with F.T. $G(y)$. Hence, by definition 7, the generalised function $g(y)$ also satisfies (30). Hence, by theorem 6, it is the F.T. of the generalised function $f(x)$. This theorem again eliminates possibilities of confusion, this time between different uses of the expression 'Fourier transform'.

2.4. Equality of a generalised function and an ordinary function in an interval

DEFINITION 8. *If $h(x)$ is an ordinary function and $f(x)$ a generalised function, and*

$$\int_{-\infty}^{\infty} f(x)\,F(x)\,\mathrm{d}x = \int_{a}^{b} h(x)\,F(x)\,\mathrm{d}x \tag{31}$$

for every good function $F(x)$ which is zero outside $a < x < b$ (here, a and b may be finite or infinite, and we assume the existence of the right-hand side of (31) as an ordinary integral for all such $F(x)$, thus imposing a restriction on the function $h(x)$ in $a < x < b$, although it need not even be defined elsewhere), then we write

$$f(x) = h(x) \quad for \quad a < x < b. \tag{32}$$

PROOF OF CONSISTENCY. The definition is consistent with the maxim that everything is equal to itself, since if $h(x)$ satisfies the condition of definition 7, then the generalised function $h(x)$ equals (in the sense of definition 8) the ordinary function $h(x)$ in *any* interval.

EXAMPLE 11. $\delta(x) = 0$ for $0 < x < \infty$ and for $-\infty < x < 0$.

PROOF. If $F(x)$ vanishes outside either of these two intervals, then $F(0) = 0$, and so by equation (10)

$$\int_{-\infty}^{\infty} \delta(x)\,F(x)\,\mathrm{d}x = 0.$$

THEOREM 12. *If $h(x)$ and its derivative $h'(x)$ are ordinary functions both satisfying the restriction imposed on $h(x)$ in definition 8, and $f(x)$ is a generalised function which equals $h(x)$ in $a < x < b$, then*

$$f'(x) = h'(x) \quad in \quad a < x < b.$$

PROOF. If $F(x)$ is a good function which is zero outside $a < x < b$, then

$$\int_{-\infty}^{\infty} f'(x)\,F(x)\,\mathrm{d}x = -\int_{-\infty}^{\infty} f(x)\,F'(x)\,\mathrm{d}x$$

$$= -\int_{a}^{b} h(x)\,F'(x)\,\mathrm{d}x = \int_{a}^{b} h'(x)\,F(x)\,\mathrm{d}x, \tag{33}$$

which proves the theorem. The assumption in the last integration by parts that $h(x)\,F(x) \to 0$ as $x \to a$ (and, similarly, as $x \to b$) is proved exactly like the corresponding assumption in the proof of theorem 10. The product must tend to some finite limit for any good function $F(x)$ vanishing outside $a < x < b$, and if $a = -\infty$ the limit is zero because $\int_a^b h(x) F(x)\, dx$ exists. If $a > -\infty$, however, it is zero because otherwise $h(x) F(x)/(x-a)$ would not tend to a finite limit, although $F(x)/(x-a)$ is itself a good function vanishing outside $a < x < b$.

EXAMPLE 12. Any repeated derivative $\delta^{(n)}(x)$ of the delta function equals o for $0 < x < \infty$ and for $-\infty < x < 0$ (by theorem 12 and example 11). It follows at once that any linear combination of the $\delta^{(n)}(x)$ similarly vanishes everywhere except at $x = 0$, which is interesting as showing what a wide variety of *different* generalised functions can all be equal at all points save one.

EXAMPLE 13. If $f(x)$, $g(x)$ are generalised functions such that $xf(x) = g(x)$, and if $g(x)$ equals an ordinary function $h(x)$ in an interval $a < x < b$ not including $x = 0$, then $f(x) = x^{-1}h(x)$ in $a < x < b$.

PROOF. If $F(x)$ is a good function which is zero outside $a < x < b$, then so is $x^{-1}F(x)$. Hence,

$$\int_{-\infty}^{\infty} f(x) F(x)\, dx = \int_{-\infty}^{\infty} xf(x)\, x^{-1}F(x)\, dx$$

$$= \int_{-\infty}^{\infty} g(x)\, x^{-1}F(x)\, dx = \int_a^b h(x)\, x^{-1}F(x)\, dx, \quad (34)$$

which proves the result. Note that, by theorem 9, the various functions $f(x)$ with $xf(x) = g(x)$ all differ by constant multiples of $\delta(x)$, which does not invalidate the result since $\delta(x) = 0$ in such an interval.

2.5. Even and odd generalised functions

DEFINITION 9. *The generalised function $f(x)$ is said to be even (or odd, respectively) if* $\int_{-\infty}^{\infty} f(x) F(x)\, dx = 0$ *for all odd (or even) good functions $F(x)$.*

EXAMPLE 14. $\delta(x)$ is even.

PROOF. $F(0) = 0$ for all odd good functions $F(x)$.

EXAMPLE 15. If $f(x)$ is an even (or odd) ordinary function satisfying the condition of definition 7, then the generalised function $f(x)$ is even (or odd).

PROOF. This follows immediately from definition 9.

THEOREM 13. *If the generalised function $f(x)$ is even (or odd, respectively), then its derivative $f'(x)$ is odd (or even), its F.T. $g(y)$ is even (or odd), while $\phi(x)f(x)$ is even (or odd) when the fairly good function $\phi(x)$ is even, and odd (or even) when $\phi(x)$ is odd.*

PROOF. This follows immediately from theorem 6 and the corresponding results for good functions.

EXAMPLE 16. $\delta^{(n)}(x)$ is even if n is even and odd if n is odd.

PROOF. This follows from example 14 and repeated application of theorem 13.

THEOREM 14. *If $f(x)$ is an even (or odd) generalised function, which equals an ordinary function $h(x)$ in the interval $a < x < b$, then*

$$f(x) = \pm h(-x) \quad in \quad -b < x < -a, \tag{35}$$

with the upper sign if $f(x)$ is even and the lower if $f(x)$ is odd.

PROOF. If $F(x)$ is zero outside $-b < x < -a$, then $F(-x)$ is zero outside $a < x < b$, and also $F(x) \mp F(-x)$ is odd (or even). Hence

$$\int_{-\infty}^{\infty} f(x)\,F(x)\,\mathrm{d}x = \pm \int_{-\infty}^{\infty} f(x)\,F(-x)\,\mathrm{d}x$$

$$= \pm \int_{a}^{b} h(x)\,F(-x)\,\mathrm{d}x = \pm \int_{-b}^{-a} h(-x)\,F(x)\,\mathrm{d}x,$$

which proves the theorem.

2.6. Limits of generalised functions

DEFINITION 10. *If $f_t(x)$ is a generalised function of x for each value of the parameter t, and $f(x)$ is another generalised function, such that, for any good function $F(x)$,*

$$\lim_{t \to c} \int_{-\infty}^{\infty} f_t(x)\,F(x)\,\mathrm{d}x = \int_{-\infty}^{\infty} f(x)\,F(x)\,\mathrm{d}x, \tag{36}$$

then we say $$\lim_{t \to c} f_t(x) = f(x). \tag{37}$$

Here, c may be finite or infinite, and t may tend to c through all real values or (when $c=\infty$) through integer values only.

THEOREM 15. *Under the conditions of definition* 10,

$$\lim_{t\to c} f'_t(x)=f'(x), \quad \lim_{t\to c} f_t(ax+b)=f(ax+b),$$

$$\lim_{t\to c} \phi(x)\,f_t(x)=\phi(x)\,f(x),$$

$$(38)$$

for any fairly good function $\phi(x)$, and

$$\lim_{t\to c} g_t(y)=g(y), \tag{39}$$

where $g_t(y)$ and $g(y)$ are the F.T.'s of $f_t(x)$ and $f(x)$.

PROOF. The proof of this remarkable theorem offers no difficulty, following closely the lines of the proof of consistency of definition 6. For example, if $F(x)$ is any good function,

$$\lim_{t\to c}\int_{-\infty}^{\infty} f'_t(x)\,F(x)\,dx = -\lim_{t\to c}\int_{-\infty}^{\infty} f_t(x)\,F'(x)\,dx$$

$$= -\int_{-\infty}^{\infty} f(x)\,F'(x)\,dx = \int_{-\infty}^{\infty} f'(x)\,F(x)\,dx, \quad (40)$$

whence by definition 10 the first result follows; and similarly with the others.

EXAMPLE 17. $\lim_{\epsilon\to 0} \epsilon\,|\,x\,|^{\epsilon-1}=2\delta(x).$

PROOF. This is obtained by differentiating the result

$$\lim_{\epsilon\to 0} |\,x\,|^{\epsilon}\operatorname{sgn} x = \operatorname{sgn} x \tag{41}$$

and using example 10 and theorem 15. To prove (41), let $F(x)$ be and good function. Then

$$\left|\int_{-\infty}^{\infty} (|\,x\,|^{\epsilon}-1)\operatorname{sgn} x F(x)\,dx\right|$$

$$\leqslant \tfrac{1}{2}\epsilon\int_{-\infty}^{\infty} |\log|\,x\,||\,(1+|\,x\,|^{\epsilon})|\,F(x)\,|\,dx = O(\epsilon) \quad (42)$$

as $\epsilon\to 0$, where the inequality $|\tanh z|\leqslant|\,z\,|$ has been applied to $z=\tfrac{1}{2}\epsilon\log|\,x\,|$.

Two specially useful kinds of limiting operation are now given their usual special names and symbols.

DEFINITION 11. *If $f_t(x)$ is a generalised function of x for each value of the parameter t, we define*

$$\frac{\partial}{\partial t} f_t(x) = \lim_{t_1 \to t} \frac{f_{t_1}(x) - f_t(x)}{t_1 - t}, \quad \sum_{t=0}^{\infty} f_t(x) = \lim_{n \to \infty} \sum_{t=0}^{n} f_t(x), \quad (43)$$

provided in each case that the limit function exists.

EXAMPLE 18. If $(\partial/\partial t) f_t(x)$ exists, then $\dfrac{\partial}{\partial t} \left\{ \dfrac{\partial}{\partial x} f_t(x) \right\}$ exists and equals $\dfrac{\partial}{\partial x} \left\{ \dfrac{\partial}{\partial t} f_t(x) \right\}$.

PROOF. By theorem 15,

$$\lim_{t_1 \to t} \frac{f'_{t_1}(x) - f'_t(x)}{t_1 - t} = \frac{\mathrm{d}}{\mathrm{d}x} \lim_{t_1 \to t} \frac{f_{t_1}(x) - f_t(x)}{t_1 - t}.$$

NOTE. Similarly, the F.T. of $(\partial/\partial t) f_t(x)$ is $(\partial/\partial t) g_t(y)$. Theorem 15 also shows that we can differentiate or take Fourier transforms of series, term by term. This fact will be continually used in chapter 5.

EXERCISE 1. Prove that

$$x^n \, \delta^{(m)}(x) = (-1)^n \frac{m!}{(m-n)!} \delta^{(m-n)}(x) \quad (m \geqslant n), \qquad 0 \quad (m < n). \tag{44}$$

Prove that the general solution of $f^{(n)}(x) = 0$ is a polynomial of degree $n - 1$. By taking Fourier transforms of this result, or otherwise, prove that the general solution of $x^n f(x) = 0$ is a linear combination of $\delta(x)$ and its first $(n - 1)$ derivatives.

EXERCISE 2. If $\phi(x)$ is any fairly good function, prove that

$$\phi(x) \, \delta(x) = \phi(0) \, \delta(x). \tag{45}$$

More generally, from the results of example 8 and exercise 1, or otherwise, prove that

$$\phi(x) \, \delta^{(m)}(x) = \sum_{n=0}^{m} (-1)^n \frac{m!}{n! \, (m-n)!} \, \phi^{(n)}(0) \, \delta^{(m-n)}(x). \tag{46}$$

EXERCISE 3. If $f(x)$ is a generalised function and $g(y)$ its F.T., find the F.T. of $x^n f(x)$.

EXERCISE 4. Prove that

$$\lim_{n \to \infty} \frac{\sin nx}{\pi x} = \delta(x). \tag{47}$$

NOTE. This is the Fourier transform of a much simpler result.

CHAPTER 3

DEFINITIONS, PROPERTIES AND FOURIER TRANSFORMS OF PARTICULAR GENERALISED FUNCTIONS

3.1. Non-integral powers

In this chapter a number of particular generalised functions are defined and studied, some for their intrinsic interest and widespread utility, and others solely for their application to the technique of asymptotic estimation of Fourier transforms described in chapter 4. We begin by defining non-integral powers, and more precisely (since non-integral powers of *negative* numbers have without further particularisation no precise meaning) the functions

$$|x|^\alpha, \quad |x|^\alpha \operatorname{sgn} x \quad \text{and} \quad x^\alpha H(x) = \tfrac{1}{2}(|x|^\alpha + |x|^\alpha \operatorname{sgn} x), \quad (1)$$

where $H(x)$, equal to 1 for $x > 0$ and 0 for $x < 0$, is Heaviside's unit function. The first of expressions (1) is an even function, the second odd, and the third (the mean of the other two) vanishes for $x < 0$.

These expressions are generalised functions of x by definition 7 when $\alpha > -1$. The differentiation rules

$$\frac{\mathrm{d}}{\mathrm{d}x}|x|^\alpha = \alpha|x|^{\alpha-1}\operatorname{sgn} x, \quad \frac{\mathrm{d}}{\mathrm{d}x}|x|^\alpha \operatorname{sgn} x = \alpha|x|^{\alpha-1}, \\ \frac{\mathrm{d}}{\mathrm{d}x}x^\alpha H(x) = \alpha x^{\alpha-1}H(x), \quad\quad\quad (2)$$

follow from theorem 10, provided that $\alpha > 0$ (so that $\alpha - 1$ as well as α exceeds -1). It is convenient to use these equations for $\alpha < 0$, repeatedly if necessary, to define the appropriate generalised functions for non-integral α less than -1.

DEFINITION 12. *If $\alpha < -1$ and is not an integer, then we define three new generalised functions as follows:*

$$\left.\begin{array}{r}|x|^\alpha \\ |x|^\alpha \operatorname{sgn} x \\ x^\alpha H(x)\end{array}\right\} = \frac{1}{(\alpha+1)(\alpha+2)\dots(\alpha+n)}\frac{\mathrm{d}^n}{\mathrm{d}x^n}\left\{\begin{array}{c}|x|^{\alpha+n}(\operatorname{sgn} x)^n \\ |x|^{\alpha+n}(\operatorname{sgn} x)^{n+1} \\ x^{\alpha+n}H(x)\end{array}\right\}, \quad (3)$$

where n is an integer such that $\alpha + n > -1$.

PROOF OF CONSISTENCY. The value of n chosen is immaterial, by equations (2). Definition 12 now extends the validity of these equations to all non-integer α. The relation (1) between the three functions, known in the first place for $\alpha > -1$, is similarly extended, by repeated differentiation, to the three new generalised functions defined by equation (3). By theorem 13, $|x|^{\alpha}$ is an even, and $|x|^{\alpha}\operatorname{sgn} x$ an odd, generalised function.

Each of the three new generalised functions of definition 12 is equal to the ordinary function of the same designation in the intervals $0 < x < \infty$ and $-\infty < x < 0$, by repeated application of theorem 12. (Note that the restriction on $h(x)$ imposed in definition 8 is satisfied by all these functions in the intervals stated, because a good function $F(x)$ which is zero outside either interval must tend to zero faster than any power of $|x|$ as $x \to 0$.) Hence, as the ordinary functions are undefined at $x = 0$, no conflict between the new and established usages is possible.

One can use definition 12 to interpret 'improper' integrals like $\int_0^a x^{\alpha} F(x)\,\mathrm{d}x$, in which α is non-integral and < -1 and $F(x)$ is a good function,* as

$$\int_{-\infty}^{\infty} \{x^{\alpha} H(x) - x^{\alpha} H(x-a)\}\, F(x)\, \mathrm{d}x, \tag{4}$$

where the second term in curly brackets is to be interpreted as a generalised function by definition 7. Integration by parts can be used to express (4) as an ordinary integral. We need the fact that, by theorem 10, if $f(x)$ is an ordinary function (here x^{α}) differentiable for $x \geqslant a$, then

$$\frac{\mathrm{d}}{\mathrm{d}x}\{f(x)\,H(x-a)\} = \frac{\mathrm{d}}{\mathrm{d}x}[\{f(x) - f(a)\}\,H(x-a)] + f(a)\,\delta(x-a)$$

$$= f'(x)\,H(x-a) + f(a)\,\delta(x-a), \tag{5}$$

which is easy to remember because of its similarity to the ordinary rule for differentiating a product. Applying this repeatedly to the

* More generally, it is sufficient that $F(x)$ be differentiable any number of times in some interval which includes $(0, a)$. This is because any such function necessarily coincides with some good function in $(0, a)$.

second term (and using equation (3) for the first), we obtain

$$x^\alpha H(x) - x^\alpha H(x-a)$$

$$= \frac{1}{(\alpha+1)(\alpha+2)\dots(\alpha+n)} \frac{d^n}{dx^n} [x^{\alpha+n}\{H(x)-H(x-a)\}]$$

$$+ \frac{a^{\alpha+1}}{\alpha+1} \delta(x-a) + \frac{a^{\alpha+2}}{(\alpha+1)(\alpha+2)} \delta'(x-a) + \dots$$

$$+ \frac{a^{\alpha+n}}{(\alpha+1)(\alpha+2)\dots(\alpha+n)} \delta^{(n-1)}(x-a), \tag{6}$$

which on substitution in (4) gives the interpretation

$$\int_0^a x^\alpha F(x)\,dx = \frac{(-1)^n}{(\alpha+1)(\alpha+2)\dots(\alpha+n)} \int_0^a x^{\alpha+n} F^{(n)}(x)\,dx$$

$$+ \frac{a^{\alpha+1}}{\alpha+1} F(a) - \frac{a^{\alpha+2}}{(\alpha+1)(\alpha+2)} F'(a) + \dots$$

$$+ \frac{(-1)^{n-1} a^{\alpha+n}}{(\alpha+1)(\alpha+2)\dots(\alpha+n)} F^{(n-1)}(a). \tag{7}$$

Equation (7) is what the ordinary formula for repeated integration by parts would give formally, if all the contributions from the lower limit (which involve $0^{\alpha+1}$, $0^{\alpha+2}$, ..., $0^{\alpha+n}$, and so are 'infinite' if n is the least integer with $\alpha+n > -1$) were omitted. For this reason the interpretation (7) was given the name 'finite part' by Hadamard, who showed that the finite part obeys many of the ordinary rules of integration. Its inclusion within the framework of generalised-function theory makes this conclusion readily intelligible.

The rest of this section is devoted to obtaining the Fourier transforms of the functions of definition 12. We need some properties of the factorial function $\alpha!$, called $\Gamma(\alpha+1)$ in the older books, and defined for $\alpha > -1$ as

$$\int_0^\infty x^\alpha e^{-x}\,dx = \alpha!, \tag{8}$$

where x^α is of course positive. The straight path of integration in (8) may be rotated about the origin through any angle $< \frac{1}{2}\pi$, provided that x^α is taken as a function regular in the right-hand half plane which is equal to the positive number x^α for real positive x; in other words, the many-valued function x^α must be made precise by taking $|\arg x| < \frac{1}{2}\pi$. Hence, if a is any complex number with positive real part,

$$\int_0^\infty x^\alpha e^{-ax} \, dx = a^{-\alpha-1} \int_0^\infty z^\alpha e^{-z} \, dz = \alpha! \, a^{-\alpha-1}, \qquad (9)$$

by the substitution $ax = z$; and in (9) we must have $|\arg a| < \frac{1}{2}\pi$, in order that $\arg z = \arg a$ shall satisfy the same condition.

To obtain the F.T. of $x^\alpha H(x)$ for $\alpha > -1$, we use the limit property

$$\lim_{t \to 0} x^\alpha e^{-xt} H(x) = x^\alpha H(x), \qquad (10)$$

which holds for $\alpha > -1$ because for any good function $F(x)$

$$\left| \int_0^\infty (1 - e^{-xt}) x^\alpha F(x) \, dx \right| \leqslant A \int_0^\infty \frac{1 - e^{-xt}}{x^{\frac{3}{2}}} \, dx = 2A \sqrt{(\pi t)} \to 0 \quad (11)$$

as $t \to 0$, where $A = \max |x^{\alpha+\frac{3}{2}} F(x)|$. Now, by theorem 11, and equation (9) above, the F.T. of $x^\alpha e^{-xt} H(x)$ is

$$\int_0^\infty x^\alpha e^{-(t+2\pi i y)x} \, dx = \alpha! \, (t + 2\pi i y)^{-\alpha-1} = \frac{d^n}{dy^n} \frac{(\alpha-n)! \, (t + 2\pi i y)^{n-\alpha-1}}{(-2\pi i)^n}. \qquad (12)$$

But, by theorem 15, the F.T. of $x^\alpha H(x)$ is the limit of (12) as $t \to 0$, and this is the nth derivative of the limit of the quotient, namely,

$$\frac{d^n}{dy^n} \frac{(\alpha-n)! \, (2\pi |y|)^{n-\alpha-1} e^{\frac{1}{2}\pi i (n-\alpha-1) \operatorname{sgn} y}}{(-2\pi i)^n}$$

$$= \{e^{-\frac{1}{2}\pi i (\alpha+1) \operatorname{sgn} y}\} \alpha! \, (2\pi |y|)^{-\alpha-1}. \qquad (13)$$

In (13) the limit of the quotient is given for $n - \alpha - 1 > 0$, when the limiting result is easy to prove, using the rule about the argument of a in (9). The nth derivative of this limit is then evaluated from definition 12, which defines as repeated derivatives the generalised functions $|y|^{-\alpha-1}$ and $|y|^{-\alpha-1} \operatorname{sgn} y$, of which, by de Moivre's theorem, expression (13) is a linear combination.

Now, by theorem 7, the F.T. of $f'(x)$ is $2\pi i y g(y)$ if the F.T. of $f(x)$ is $g(y)$. Applied as a check to (13), this says that the F.T. of $\alpha x^{\alpha-1} H(x)$ is

$$\{e^{-\frac{1}{2}\pi i \alpha \operatorname{sgn} y}\} \alpha! \, (2\pi |y|)^{-\alpha}, \qquad (14)$$

since $iy = e^{\frac{1}{2}\pi i \, \text{sgn} \, y} |y|$. Equation (14) shows that the expression (13) for the F.T. of $x^\alpha H(x)$ is valid for $\alpha - 1$ if it is valid for α, and so extends it by induction to all values of α, which is necessary because the proof given, using the integral expression for the factorial function, applies only when $\alpha > -1$.

Since

$$\left.\begin{array}{l} |x|^\alpha = x^\alpha H(x) + (-x)^\alpha H(-x), \\ |x|^\alpha \, \text{sgn} \, x = x^\alpha H(x) - (-x)^\alpha H(-x), \end{array}\right\} \tag{15}$$

and by theorem 7 the F.T. of $(-x)^\alpha H(-x)$ is obtained by changing the sign of y in (13), we deduce that the F.T. of $|x|^\alpha$ is

$$\{2 \cos \tfrac{1}{2}\pi(\alpha + 1)\} \alpha! (2\pi |y|)^{-\alpha - 1} \tag{16}$$

(which is even, in agreement with theorem 13), and that the F.T. of $|x|^\alpha \, \text{sgn} \, x$ is

$$\{-2i \sin \tfrac{1}{2}\pi(\alpha + 1)\} \alpha! (2\pi |y|)^{-\alpha - 1} \, \text{sgn} \, y \tag{17}$$

(which, similarly, is odd). Conversely, if (16) and (17) had been derived first, (13) would follow from them by equation (1).

EXERCISE 5. Show that

$$\int_0^1 \frac{dx}{x^{\frac{5}{2}}(1 + x)} = \tfrac{1}{2}\pi + \tfrac{4}{3}.$$

EXERCISE 6. Prove that the equation $xf(x) = |x|^\alpha$ is satisfied by $f(x) = |x|^{\alpha - 1} \, \text{sgn} \, x$. Are there any other solutions?

EXERCISE 7. Check that Fourier's inversion theorem (see theorem 7) is satisfied both by $|x|^\alpha$ and by $|x|^\alpha \, \text{sgn} \, x$.

NOTE. The standard formula

$$\alpha! (-\alpha - 1)! = -\pi \, \text{cosec} \, \pi\alpha \tag{18}$$

will be required.

3.2. Non-integral powers multiplied by logarithms

DEFINITION 13.

$$\left.\begin{array}{l} |x|^\alpha \log |x| = \dfrac{\partial}{\partial \alpha} |x|^\alpha, \quad |x|^\alpha \log |x| \, \text{sgn} \, x = \dfrac{\partial}{\partial \alpha} (|x|^\alpha \, \text{sgn} \, x), \\[2mm] x^\alpha \log x H(x) = \dfrac{\partial}{\partial \alpha} \{x^\alpha H(x)\}. \end{array}\right\} \tag{19}$$

NOTE. These equations are of course true in the ordinary sense for $\alpha > -1$. For non-integral $\alpha < -1$, they may be taken as defining the generalised functions on the left-hand sides; the derivatives with respect to α (in the sense of definition 11) then exist by repeated application of the result of example 18. This example shows also that the ordinary rules of differentiation apply to these functions; for example,

$$\frac{d}{dx}(\,|x|^\alpha \log|x|\,) = \frac{\partial}{\partial x}\frac{\partial}{\partial \alpha}\,|x|^\alpha = \frac{\partial}{\partial \alpha}\frac{\partial}{\partial x}\,|x|^\alpha = \frac{\partial}{\partial \alpha}(\alpha\,|x|^{\alpha-1}\,\mathrm{sgn}\,x)$$

$$= |x|^{\alpha-1}\mathrm{sgn}\,x + \alpha\,|x|^{\alpha-1}\log|x|\,\mathrm{sgn}\,x, \qquad (20)$$

as one would obtain for positive α by direct differentiation.

The F.T. of $|x|^\alpha \log|x|$, by the note following example 18, and by equation (16), is

$$\frac{\partial}{\partial \alpha}[\{2\cos \tfrac{1}{2}\pi(\alpha+1)\}\alpha!\,(2\pi\,|y|)^{-\alpha-1}]$$

$$= \{2\cos\tfrac{1}{2}\pi(\alpha+1)\}\alpha!\,(2\pi\,|y|)^{-\alpha-1}$$

$$\times \{-\log(2\pi\,|y|) + \psi(\alpha) - \tfrac{1}{2}\pi\tan\tfrac{1}{2}\pi(\alpha+1)\}, \quad (21)$$

where
$$\psi(\alpha) = \frac{d}{d\alpha}\log(\alpha!). \qquad (22)$$

Similarly, that of $|x|^\alpha \log|x|\,\mathrm{sgn}\,x$ is

$$\{-2i\sin\tfrac{1}{2}\pi(\alpha+1)\}\alpha!\,(2\pi\,|y|)^{-\alpha-1}\mathrm{sgn}\,y$$

$$\times \{-\log(2\pi\,|y|) + \psi(\alpha) + \tfrac{1}{2}\pi\cot\tfrac{1}{2}\pi(\alpha+1)\}, \quad (23)$$

and that of $x^\alpha \log x\, H(x)$ is

$$\{e^{-\frac{1}{2}\pi i(\alpha+1)\,\mathrm{sgn}\,y}\}\alpha!\,(2\pi\,|y|)^{-\alpha-1}\{-\log(2\pi\,|y|) + \psi(\alpha) - \tfrac{1}{2}\pi i\,\mathrm{sgn}\,y\}.$$

$$(24)$$

3.3. Integral powers

Throughout the rest of this chapter, n signifies any integer $\geqslant 0$ and m any integer > 0. By the second and third parts of theorem 7, if $f(x)$ has F.T. $g(y)$ then

$$x^n f(x) \quad \text{has F.T.} \quad (-2\pi i)^{-n} g^{(n)}(y). \qquad (25)$$

(This is the solution to exercise 3.)

Hence, by example 7, the F.T. of x^n is $(-2\pi i)^{-n}\delta^{(n)}(y)$. This, as theorem 15 requires, is actually the limit as $\alpha \to n$ of expression (16) when n is even, namely

$$\lim_{\alpha \to n} \left[\{2\cos\tfrac{1}{2}\pi(\alpha+1)\} \frac{\alpha!}{(-\alpha)(-\alpha+1)\dots(-\alpha+n-1)} \right.$$

$$\left. \times \frac{1}{(2\pi)^{\alpha+1}} \frac{d^n}{dy^n} \frac{1}{|y|^{1+\alpha-n}} \right]$$

$$= \left\{ \lim_{\alpha \to n} \frac{2\cos\tfrac{1}{2}\pi(\alpha+1)}{n-\alpha} \right\} \frac{1}{(2\pi)^{n+1}} \frac{d^n}{dy^n} \left\{ \lim_{\alpha \to n} \frac{n-\alpha}{|y|^{1+\alpha-n}} \right\}$$

$$= \pi i^n \frac{1}{(2\pi)^{n+1}} \frac{d^n}{dy^n} \{2\delta(y)\} = (-2\pi i)^{-n}\delta^{(n)}(y), \qquad (26)$$

where theorem 15 and example 17 have been used. Similarly, it is the limit as $\alpha \to n$ of (17) when n is odd.

We next define negative integral powers of x; the only problem is how to define x^{-1}, the others being deduced from it by differentiation.

DEFINITION 14. *x^{-1} is the odd generalised function satisfying $xf(x)=1$; and*

$$x^{-m} = \frac{(-1)^{m-1}}{(m-1)!} \frac{d^{m-1}}{dx^{m-1}} (x^{-1}). \qquad (27)$$

PROOF OF CONSISTENCY. The equation $xf(x)=1$ has such a solution, since $f(x)=d\log|x|/dx$ is an odd generalised function by theorem 13 and example 15, and it satisfies

$$xf(x) = d(x\log|x|)/dx - \log|x| = 1. \qquad (28)$$

By theorem 9, the general solution is $f(x)+C\delta(x)$, but this is odd only for $C=0$, by example 14. Hence the definition specifies x^{-1} uniquely, and theorem 12 shows that it equals the ordinary function x^{-1} in $0<x<\infty$ and in $-\infty<x<0$. Note also that x^{-m} has a similar property and (by theorem 13) is even when m is even and odd when m is odd.

Now, if $\operatorname{sgn} x$ has F.T. $g(y)$, then by example 10, theorem 7 and example 7, $2\pi i y g(y)=2$. But $g(y)$ is odd by theorem 13, and so (by definition 14) it is $(\pi i)^{-1}y^{-1}$. Hence also, by equation (25), the F.T. of $x^n \operatorname{sgn} x$ is

$$(-2\pi i)^{-n}(\pi i)^{-1}\frac{d^n}{dy^n}(y^{-1}) = 2(n!)(2\pi i y)^{-n-1}. \qquad (29)$$

Now, when n is even, $x^n \operatorname{sgn} x = \lim_{\alpha \to n} |x|^\alpha \operatorname{sgn} x$, and so, by comparing the limit of (17) as $\alpha \to n$ with (29) (and using theorem 15), we see that

$$\lim_{\alpha \to n} |y|^{-\alpha-1} \operatorname{sgn} y = y^{-n-1}. \tag{30}$$

When n is odd, $x^n \operatorname{sgn} x = \lim_{\alpha \to n} |x|^\alpha$, and so (29) must be compared with the limit of (16), and we infer that in this case

$$\lim_{\alpha \to n} |y|^{-\alpha-1} = y^{-n-1}. \tag{31}$$

These properties are satisfactory features of definition 14.

Definition 14 is also closely connected with Cauchy's 'principal value' of an integral whose integrand has an inverse-first-power singularity. In the present theory, if $F(x)$ is a good function and $a < 0 < b$, the integral $\int_a^b x^{-1} F(x)\, dx$ must be interpreted as

$$\int_{-\infty}^\infty \{x^{-1} - x^{-1} H(x-b) - x^{-1} H(a-x)\} F(x)\, dx. \tag{32}$$

But, by equation (5),

$$x^{-1} - x^{-1} H(x-b) - x^{-1} H(a-x)$$

$$= \frac{d}{dx} [\log |x| \{1 - H(x-b) - H(a-x)\}]$$

$$+ (\log b)\, \delta(x-b) - (\log |a|)\, \delta(x-a), \tag{33}$$

which substituted in (32) gives the interpretation

$$\int_a^b x^{-1} F(x)\, dx = - \int_a^b (\log |x|) F'(x)\, dx + F(b) \log b - F(a) \log |a|. \tag{34}$$

This may be compared with the 'Cauchy principal value', defined as

$$\lim_{\epsilon \to 0} \left(\int_a^{-\epsilon} + \int_\epsilon^b \right) x^{-1} F(x)\, dx = \lim_{\epsilon \to 0} \left\{ - \left(\int_a^{-\epsilon} + \int_\epsilon^b \right) (\log |x|) F'(x)\, dx \right.$$

$$\left. + F(b) \log b - F(\epsilon) \log \epsilon + F(-\epsilon) \log \epsilon - F(a) \log |a| \right\}, \tag{35}$$

which is seen to be identical with (34).

In the same way, integrals involving other inverse-integral-power singularities can be interpreted by definition 14, and this interpretation also has been anticipated by a number of writers. We obtain as above (but with repeated integration by parts)

$$\int_a^b x^{-m}F(x)\,\mathrm{d}x = \frac{1}{(m-1)!}\int_a^b x^{-1}F^{(m-1)}(x)\,\mathrm{d}x - \frac{b^{1-m}F(b)-a^{1-m}F(a)}{m-1}$$

$$-\frac{b^{2-m}F'(b)-a^{2-m}F'(a)}{(m-1)(m-2)} - \cdots - \frac{b^{-1}F^{(m-2)}(b)-a^{-1}F^{(m-2)}(a)}{(m-1)(m-2)\ldots 1}, \quad (36)$$

where the Cauchy principal value itself has been left uninterpreted because most readers will be familiar with these integrals. The interpretations (36) and (34) are valid (see the footnote to equation (4)) if $F(x)$ is any function differentiable any number of times in an interval including (a, b).

We have treated x^n, $x^n \operatorname{sgn} x$ and x^{-m}; more serious difficulties are presented by $x^{-m}\operatorname{sgn} x$. These difficulties are already fully present in the case $m = 1$, that is, for the function $|x|^{-1}$. The limit of the generalised function $|x|^{\epsilon-1}$ as $\epsilon \to 0$ does not exist, as example 17 clearly shows. Again, all the solutions of the equation

$$xf(x) = \operatorname{sgn} x$$

are even, so that the method of definition 14 cannot be used to pick out one of them. Particular solutions *can* be specified, but they do not obey the manipulation rules which one expects of a function called $|x|^{-1}$.

EXAMPLE 19. If $f(x) = \mathrm{d}(\log|x|\operatorname{sgn} x)/\mathrm{d}x$, then $xf(x) = \operatorname{sgn} x$, but $f(ax) \neq |a|^{-1}f(x)$.

PROOF. The first part is proved as in equation (28); also

$$f(ax) = \frac{\mathrm{d}\{(\log|x| + \log|a|)\operatorname{sgn} a \operatorname{sgn} x\}}{a\,\mathrm{d}x}$$

$$= \frac{1}{|a|}\frac{\mathrm{d}}{\mathrm{d}x}(\log|x|\operatorname{sgn} x + \log|a|\operatorname{sgn} x)$$

$$= \frac{1}{|a|}\{f(x) + 2\log|a|\,\delta(x)\}. \quad (37)$$

The only satisfactory definition is one which admits the indeterminacy, in the same way as does the definition of the 'indefinite integral'.

DEFINITION 15. *The symbol $|x|^{-1}$ will stand for any generalised function $f(x)$ such that $xf(x) = \text{sgn } x$. The symbol $x^{-m} \text{sgn } x$ will stand for $(-1)^{m-1}/(m-1)!$ times the $(m-1)$th derivative of any of these functions.*

Thus (by theorem 9 and example 19) $|x|^{-1}$ can be written as

$$\frac{\mathrm{d}}{\mathrm{d}x}(\log|x|\,\text{sgn } x) + C\delta(x) = \frac{\mathrm{d}}{\mathrm{d}x}\{(\log|x|+C)\,\text{sgn } x\}, \qquad (38)$$

and $x^{-m}\text{sgn } x$ as

$$\frac{(-1)^{m-1}}{(m-1)!}\frac{\mathrm{d}^m}{\mathrm{d}x^m}\{(\log|x|+C)\,\text{sgn } x\}$$
$$= \frac{(-1)^{m-1}}{(m-1)!}\frac{\mathrm{d}^m}{\mathrm{d}x^m}(\log|x|\,\text{sgn } x) + C\delta^{(m-1)}(x), \quad (39)$$

where C is an arbitrary constant in each expression (not the same in each). Study of example 19 shows that, with C arbitrary in this sense, definition 15 gives $|ax|^{-1} = |a|^{-1}|x|^{-1}$.

Again, it is only with C arbitrary in this sense that the relation $x(x^{-m}\text{sgn } x) = x^{-(m-1)}\text{sgn } x$ holds.

EXAMPLE 20. $\lim\limits_{\epsilon\to 0}\{|x|^{\epsilon-1} - 2\epsilon^{-1}\delta(x)\} = |x|^{-1}$.

PROOF. An equation like this means that the limit exists and equals one of the values of $|x|^{-1}$. It is obtained by differentiating the result

$$\lim_{\epsilon\to 0}\frac{(|x|^{\epsilon}-1)\,\text{sgn } x}{\epsilon} = \log|x|\,\text{sgn } x. \qquad (40)$$

EXAMPLE 21. The F.T. of $\log|x|$ is $-\frac{1}{2}|y|^{-1}$.

PROOF. Since $\log|x| = \lim\limits_{\epsilon\to 0}(1 - |x|^{-\epsilon})/\epsilon$, its F.T. (by theorem 15 and equation (16)) is

$$\lim_{\epsilon\to 0}\{\delta(y) - (2\sin\tfrac{1}{2}\pi\epsilon)(-\epsilon)!(2\pi|y|)^{\epsilon-1}\}/\epsilon$$
$$= -\tfrac{1}{2}\lim_{\epsilon\to 0}\{[1 + \epsilon\{\log(2\pi) - \psi(0)\} + O(\epsilon^2)]|y|^{\epsilon-1} - 2\epsilon^{-1}\delta(y)\}$$
$$= -\tfrac{1}{2}\lim_{\epsilon\to 0}\{|y|^{\epsilon-1} - 2\epsilon^{-1}\delta(y)\} - \{\log(2\pi) - \psi(0)\}\delta(y), \qquad (41)$$

by example 17. Expression (41) is $(-\frac{1}{2})$ times one of the values of $|y|^{-1}$ (though a different one from that in example 20).

Conversely, the F.T. of $|x|^{-1}$ is

$$-2(\log|y|+C), \tag{42}$$

where the arbitrary constant C is present because $|x|^{-1}$ contains an arbitrary multiple of $\delta(x)$. It follows from (42) by definition 15 and theorem 7 that the F.T. of $x^{-m}\operatorname{sgn}x$ is

$$-2\frac{(-2\pi iy)^{m-1}}{(m-1)!}(\log|y|+C). \tag{43}$$

Note finally that, by theorem 12 and example 12, all the determinations of $x^{-m}\operatorname{sgn}x$ are equal in the intervals $-\infty<x<0$ and $0<x<\infty$, and equal to the ordinary function $x^{-m}\operatorname{sgn}x$ in these intervals.

3.4. Integral powers multiplied by logarithms

The generalised function $x^n\log|x|$ exists by definition 7, and its F.T., by example 21 and equation (25), is

$$-\pi i\frac{n!}{(2\pi iy)^{n+1}}\operatorname{sgn}y, \tag{44}$$

a result which can be checked against equation (43) and Fourier's inversion theorem.

The generalised function $x^{-m}\log|x|$ requires definition, but like x^{-m} itself it presents no difficulty. In fact, if m is even, $|x|^\alpha\log|x|$ tends to a limit as $\alpha\to-m$, as is proved by the fact that its F.T. (expression (21)) can be written as

$$\frac{\pi}{(-\alpha-1)!\cos\frac12\pi\alpha}(2\pi|y|)^{-\alpha-1}\{-\log(2\pi|y|)+\psi(-\alpha-1)$$
$$-\pi\cot\pi\alpha+\tfrac12\pi\cot\tfrac12\pi\alpha\} \tag{45}$$

(where equation (18) and its logarithmic derivative have been used to put (45) into a convenient form), and that this tends to a limit

$$\pi i\frac{(-2\pi iy)^{m-1}}{(m-1)!}\operatorname{sgn}y\{\log(2\pi|y|)-\psi(m-1)\} \tag{46}$$

as $\alpha\to-m$. Similarly, we may infer that $|x|^\alpha\log|x|\operatorname{sgn}x$ tends to a limit $\alpha\to-m$ if m is odd, from the fact that its F.T. (23) tends to a limit, which again can be thrown into the form (46). These facts make the following definition appropriate.

DEFINITION 16. *The generalised function $x^{-m}\log|x|$ is the function whose F.T. is (46), so that it is the limit as $\alpha \to -m$ of $|x|^\alpha \log|x|$ if m is even and of $|x|^\alpha \log|x|\,\mathrm{sgn}\,x$ if m is odd.*

Fourier's inversion theorem applied to definition 16 gives us that the F.T. of $x^n \log|x|\,\mathrm{sgn}\,x$ is

$$-2\frac{n!}{(2\pi iy)^{n+1}}\{\log(2\pi|y|)-\psi(n)\}. \tag{47}$$

In connexion with (46) and (47) the reader may like to be reminded that

$$\psi(n) = -\gamma + 1 + \frac{1}{2} + \frac{1}{3} + \ldots + \frac{1}{n}, \tag{48}$$

where $\gamma = -\psi(0) = 0.5772$ is Euler's constant.

Finally, we come to the function $x^{-m}\log|x|\,\mathrm{sgn}\,x$, which like $x^{-m}\,\mathrm{sgn}\,x$ presents more serious difficulties, and involves a certain indeterminacy. It is most expeditiously approached from the special case $n=0$ of (47); the F.T. of $\log|x|\,\mathrm{sgn}\,x$ is

$$-\frac{1}{\pi iy}\{\log(2\pi|y|)+\gamma\} = -\frac{1}{2\pi i}\frac{d}{dy}\{\log(2\pi|y|)+\gamma\}^2. \tag{49}$$

If now $f(x)$ is taken as the function whose F.T. is $\{\log(2\pi|y|)+\gamma\}^2$, it follows from (49) and (25) that

$$xf(x) = \log|x|\,\mathrm{sgn}\,x. \tag{50}$$

However, by theorem 9, the general solution of equation (50) is $f(x) + C\delta(x)$, and there is no way of selecting one solution as more suitable than the others (for example, all are even) and, indeed, if C is not left arbitrary, the ordinary rules of manipulation cannot be applied to this function and those derived from it by differentiation.

DEFINITION 17. *The symbol $|x|^{-1}\log|x|$ will stand for any generalised function $f(x)$ which satisfies equation (50). The symbol $x^{-m}\log|x|\,\mathrm{sgn}\,x$ will stand for*

$$\frac{(-1)^{m-1}}{(m-1)!}f^{(m-1)}(x) + \left(1 + \frac{1}{2} + \frac{1}{3} + \ldots + \frac{1}{m-1}\right)x^{-m}\,\mathrm{sgn}\,x, \tag{51}$$

where $f(x)$ is any generalised function which satisfies (50).

NOTE. (51) is the equation relating the corresponding ordinary functions.

Under definition 17, the F.T. of $|x|^{-1}\log|x|$ is

$$\{\log(2\pi|y|)+\gamma\}^2+C, \tag{52}$$

where C is arbitrary, and that of $x^{-m}\log|x|\,\mathrm{sgn}\,x$ can be thrown into the form

$$\frac{(-2\pi iy)^{m-1}}{(m-1)!}[\{\log(2\pi|y|)-\psi(m-1)\}^2+C], \tag{53}$$

where C again is arbitrary (and not in general the same in each formula).

3.5. Summary of Fourier transform results

The complete set of Fourier transform results for the elementary functions possessing algebraic or algebraico-logarithmic singularities at $x=0$, which have been derived in the foregoing sections, are collected for easy reference in table 1. By use of this table it is possible, as will be seen in chapter 4, to write down the asymptotic behaviour of the F.T. of a function as $|y|\to\infty$ by inspecting its singularities and adding up the F.T.'s of the elementary functions which have the same singularities. As these singularities are not always at the origin, the formula which expresses the F.T. of $f(ax+b)$ in terms of that of $f(x)$ (theorem 7) must frequently be used in conjunction with table 1 and accordingly has been written under it.

The kind of singularity occurring in most applications consists of some linear combination of those in table 1. Very occasionally, however, terms involving higher powers of $\log|x|$ are present. The reader should be able to derive the F.T. of a function involving $(\log|x|)^2$ or higher powers by the methods of this chapter (for example, in §3.2, further differentiations with respect to α would be necessary); one set of results involving $(\log|x|)^2$ can in fact be obtained directly by applying Fourier's inversion theorem to the result (53).

Table 1 can also be used to find directly the Fourier transform of any rational function. By a familiar theorem in algebra any rational function can be expressed 'in partial fractions', or more precisely as a linear combination of integral powers x^n and negative integral

Table I

	1	sgn x	H(x)
$\lvert x\rvert^\alpha$	$\{2\cos\tfrac12\pi(\alpha+1)\}\,\alpha!(2\pi\lvert y\rvert)^{-\alpha-1}$	$\{-2i\sin\tfrac12\pi(\alpha+1)\}\,\alpha!(2\pi\lvert y\rvert)^{-\alpha-1}\,\mathrm{sgn}\,y$	$\{e^{-\frac12\pi i(\alpha+1)\,\mathrm{sgn}\,y}\}\,\alpha!(2\pi\lvert y\rvert)^{-\alpha-1}$
x^n	$(-2\pi i)^{-n}\,\delta^{(n)}(y)$	$2(n!)\,(2\pi i y)^{-n-1}$	$(-2\pi i)^{-n}\left\{\tfrac12\delta^{(n)}(y)+\dfrac{(-1)^n\,n!}{2\pi i y^{n+1}}\right\}$
x^{-m}	$-\pi i\,\dfrac{(-2\pi i y)^{m-1}}{(m-1)!}\,\mathrm{sgn}\,y$	$-2\,\dfrac{(-2\pi i y)^{m-1}}{(m-1)!}\,(\log\lvert y\rvert+C)$	$-\dfrac{(-2\pi i y)^{m-1}}{(m-1)!}\times\{\tfrac12\pi i\,\mathrm{sgn}\,y+\log\lvert y\rvert+C\}$
$\lvert x\rvert^\alpha\log\lvert x\rvert$	$\{2\cos\tfrac12\pi(\alpha+1)\}\,\alpha!(2\pi\lvert y\rvert)^{-\alpha-1}$ $\times\{-\log(2\pi\lvert y\rvert)+\psi(\alpha)-\tfrac12\pi\tan\tfrac12\pi(\alpha+1)\}$	$\{-2i\sin\tfrac12\pi(\alpha+1)\}\,\alpha!(2\pi\lvert y\rvert)^{-\alpha-1}\,\mathrm{sgn}\,y$ $\times\{-\log(2\pi\lvert y\rvert)+\psi(\alpha)+\tfrac12\pi\cot\tfrac12\pi(\alpha+1)\}$	$\{e^{-\frac12\pi i(\alpha+1)\,\mathrm{sgn}\,y}\}\,\alpha!(2\pi\lvert y\rvert)^{-\alpha-1}$ $\times\{-\log(2\pi\lvert y\rvert)+\psi(\alpha)-\tfrac12\pi i\,\mathrm{sgn}\,y\}$
$x^n\log\lvert x\rvert$	$-\pi i\,\dfrac{n!}{(2\pi i y)^{n+1}}$	$-2\,\dfrac{n!}{(2\pi i y)^{n+1}}\{\log(2\pi\lvert y\rvert)-\psi(n)\}$	$-\dfrac{n!}{(2\pi i y)^{n+1}}$ $\times\{\tfrac12\pi i\,\mathrm{sgn}\,y+\log(2\pi\lvert y\rvert)-\psi(n)\}$
$x^{-m}\log\lvert x\rvert$	$\pi i\,\dfrac{(-2\pi i y)^{m-1}}{(m-1)!}\,\mathrm{sgn}\,y$ $\times\{\log(2\pi\lvert y\rvert)-\psi(m-1)\}$	$\dfrac{(-2\pi i y)^{m-1}}{(m-1)!}$ $\times[\{\log(2\pi\lvert y\rvert)-\psi(m-1)\}^2+C]$	$\dfrac{(-2\pi i y)^{m-1}}{(m-1)!}\,[\tfrac12\{\tfrac12\pi i\,\mathrm{sgn}\,y$ $+\log(2\pi\lvert y\rvert)-\psi(m-1)\}^2+C]$

Table I. Each entry is the Fourier transform of the product of the expression at the head of its row with the expression at the head of its column. Thus, the F.T. of $x^n\,\mathrm{sgn}\,x$ is $2(n!)\,(2\pi i y)^{-n-1}$. In table I, as well as throughout chapter 3, α stands for any real number not an integer, n for any integer ≥ 0, m for any integer > 0, and C for an arbitrary constant which is present because $x^{-m}\,\mathrm{sgn}\,x$ and $x^{-m}\log\lvert x\rvert\,\mathrm{sgn}\,x$ are both indeterminate to the extent of an arbitrary multiple of $\delta^{(m-1)}(x)$. The table is specially valuable when used together with the fact that, if the F.T. of $f(x)$ is $g(y)$, then the F.T. of $f(ax+b)$ is $\dfrac{1}{\lvert a\rvert}\,e^{2\pi i b y/a}\,g\!\left(\dfrac{y}{a}\right).$ (Conversely, the F.T. of $e^{ikx}f(x)$ is $g\!\left(y-\dfrac{k}{2\pi}\right).$)

powers $(x-c)^{-m}$ for different real and complex values of c. The F.T. of x^n, and of $(x-c)^{-m}$ for real c, can be read off from table 1. For complex c we need the following additional result.

EXAMPLE 22. The F.T. of $\{x-(c_1+ic_2)\}^{-m}$ for $c_2 \neq 0$ is

$$2\pi i H(-c_2 y)\operatorname{sgn} c_2 \frac{(-2\pi iy)^{m-1}}{(m-1)!} e^{-2\pi iy(c_1+ic_2)}. \tag{54}$$

PROOF. The result (12), with $\alpha = m-1$ and $t = 2\pi|c_2|$, says that the F.T. of

$$f(x) = (2\pi i)^m \frac{x^{m-1}e^{-2\pi|c_2|x}H(x)}{(m-1)!} \tag{55}$$

is $g(y) = (y - i|c_2|)^{-m}$, whence by Fourier's inversion theorem the F.T. of

$$(x-ic_2)^{-m} = (\operatorname{sgn} c_2)^m g(x \operatorname{sgn} c_2) \tag{56}$$

is

$$(\operatorname{sgn} c_2)^m f(-y \operatorname{sgn} c_2) = 2\pi i \operatorname{sgn} c_2 \frac{(-2\pi iy)^{m-1}}{(m-1)!} e^{2\pi c_2 y} H(-yc_2). \tag{57}$$

Hence, by theorem 7, the F.T. of $\{x-(c_1+ic_2)\}^{-m}$ is $e^{-2\pi ic_1 y}$ times (57), as (54) in fact states.

It is worth noting that expression (54) is $2H(-c_2 y)$ times what is got by inferring (incorrectly) the F.T. of $\{x-(c_1+ic_2)\}^{-m}$ from that of x^{-m} by means of the result quoted at the bottom of table 1 (which is valid only for real a and b, and therefore applicable only for $c_2 = 0$). It follows that the F.T. of $(x-c_1)^{-m}$ is not the limit of that of $\{x-(c_1+ic_2)\}^{-m}$ as $c_2 \to 0$ either from above or below, but that (since $H(-c_2 y)+H(c_2 y)=1$) it is half the sum of the two limits. (This point emerges clearly also from a contour-integration approach to the evaluation of Fourier integrals.)

We end this section with an example on finding the F.T. of a rational function.

EXAMPLE 23. The F.T. of $f(x) = x^5/(x^4 - 1)$ is

$$g(y) = \frac{i\delta'(y)}{2\pi} + \tfrac{1}{2}\pi i(e^{-2\pi|y|} - \cos 2\pi y)\operatorname{sgn} y. \tag{58}$$

PROOF. In partial fractions,

$$f(x) = x + \frac{\tfrac{1}{4}}{x+1} + \frac{\tfrac{1}{4}}{x-1} - \frac{\tfrac{1}{4}}{x+i} - \frac{\tfrac{1}{4}}{x-i}, \tag{59}$$

whence, by table 1 and example 22,

$$g(y) = -\frac{1}{2\pi i}\delta'(y) + \tfrac{1}{4}(e^{2\pi i y} + e^{-2\pi i y})(-\pi i \operatorname{sgn} y)$$
$$- \tfrac{1}{4}(2\pi i)\{-H(y)\,e^{-2\pi y} + H(-y)\,e^{2\pi y}\}, \quad (60)$$

which on simplification takes the form given.

EXERCISE 8. Show that

$$\int_{-\infty}^{\infty} x^{-4}\,e^{-x^2}\,dx = \tfrac{4}{3}\sqrt{\pi}.$$

EXERCISE 9. Find the F.T. of

$$(1-x)^{-\frac{3}{2}}\log(1-x)\,H(1-x).$$

EXERCISE 10. Find the F.T. of

(i) $(x^2 + 5x + 4)^{-2}$, (ii) $(x^2 + 2x + 5)^{-2}$.

CHAPTER 4

THE ASYMPTOTIC ESTIMATION OF FOURIER TRANSFORMS

4.1. The Riemann–Lebesgue lemma

An asymptotic expression for a function is an expression as the sum of a simpler function and of a remainder which tends to zero at infinity, or (more generally) which tends to zero after multiplication by some power. When we do not know the Fourier transform of a given function, it is convenient to possess at least an asymptotic expression for it. In this chapter we develop a method which leads quickly to such an asymptotic expression for most functions occurring in applications.

The method involves writing the given function, say $f(x)$, as the sum of a simpler function $F(x)$, whose F.T. $G(y)$ we know, and of a remainder $f_R(x)$, whose F.T. $g_R(y)$ tends to zero, or (more generally) is such that the F.T. $(2\pi i y)^N g_R(y)$ of its Nth derivative $f_R^{(N)}(x)$ tends to zero. Then the F.T. of $f(x)$, say $g(y)$, satisfies

$$g(y) = G(y) + g_R(y) = G(y) + o(|y|^{-N}) \tag{1}$$

as $|y| \to \infty$.

To develop such a method, we need a simple technique for recognising functions whose Fourier transforms must tend to zero as $|y| \to \infty$. The following theorem is the classical result which does this for ordinary functions.

THEOREM 16 (The Riemann–Lebesgue lemma). *If $f(x)$ is an ordinary function absolutely integrable from $-\infty$ to ∞, and $g(y)$ is its F.T., then $g(y) \to 0$ as $|y| \to \infty$.*

PROOF. This theorem is proved in standard books on analysis; we remind the reader that one writes

$$g(y) = \int_{-\infty}^{\infty} f(x) e^{-2\pi i x y} \, dx = -\int_{-\infty}^{\infty} f\left(x + \frac{1}{2y}\right) e^{-2\pi i x y} \, dx \tag{2}$$

by a simple substitution, whence

$$|g(y)| = \left| \frac{1}{2} \int_{-\infty}^{\infty} \left\{ f(x) - f\left(x + \frac{1}{2y}\right) \right\} e^{-2\pi \mathrm{1}xy} \, dx \right|$$

$$\leqslant \frac{1}{2} \int_{-\infty}^{\infty} \left| f(x) - f\left(x + \frac{1}{2y}\right) \right| dx, \tag{3}$$

which tends to 0 as $|y| \to \infty$ by a fundamental theorem of integration.

To use theorem 16 one must be able to recognise absolute integrability in commonly occurring functions. The most useful test is as follows.

EXAMPLE 24. If $f(x)$ is continuous except at $x = x_1$, $x = x_2$, ..., $x = x_M$, and if

$$f(x) = O(|x - x_m|^{\beta_m}) \quad \text{as} \quad x \to x_m, \quad \text{where} \quad \beta_m > -1 \tag{4}$$

for $m = 1$ to M, and

$$f(x) = O(|x|^{\beta_0}) \quad \text{as} \quad |x| \to \infty, \quad \text{where} \quad \beta_0 < -1, \tag{5}$$

then $f(x)$ is absolutely integrable from $-\infty$ to ∞ and therefore its F.T. $g(y) \to 0$ as $|y| \to \infty$.

PROOF is immediate from the integrability properties of the comparison functions.

EXAMPLE 25. The function

$$f(x) = |x^4 - 1|^{-\frac{1}{2}} \operatorname{sgn} x \tag{6}$$

satisfies the conditions of example 24, with $x_1 = -1$, $x_2 = 0$, $x_3 = 1$, $\beta_1 = -\frac{1}{2}$, $\beta_2 = 0$, $\beta_3 = -\frac{1}{2}$ and $\beta_0 = -2$. Hence its F.T. $g(y) \to 0$ as $|y| \to \infty$.

4.2. Generalisations of the Riemann–Lebesgue lemma

We need results similar to theorem 16 for generalised functions. We begin with some rather trivial definitions.

DEFINITION 18. *If $g(y)$ is a generalised function, then any statement like*

$$g(y) \to 0, \quad g(y) = O\{h(y)\}, \quad \text{or} \quad g(y) = o\{h(y)\}, \tag{7}$$

as $y \to c$ (or as $|y| \to \infty$), means that $g(y)$ is equal in some interval including $y = c$ (or in some interval $|y| > R$, respectively) to an ordinary function $g_1(y)$ satisfying the stated condition.

EXAMPLE 26. $\delta(y) + y^{-1} \to 0$ as $|y| \to \infty$.

DEFINITION 19. *If $f(x)$ is a generalised function which equals an ordinary function $f_1(x)$ in the interval $a < x < b$, and $f_1(x)$ is absolutely integrable in the interval (a, b), then we say that $f(x)$ is absolutely integrable in (a, b).*

EXAMPLE 27. $\delta(x)$ is absolutely integrable in $(0, \infty)$ and $(-\infty, 0)$, but not in $(-\infty, \infty)$.

THEOREM 17. *If a generalised function $f(x)$ is absolutely integrable in $(-\infty, \infty)$ and $g(y)$ is its F.T., then $g(y) \to 0$ as $|y| \to \infty$.*

PROOF. This, by definitions 18 and 19, is just a rewritten version of the Riemann–Lebesgue lemma.

Now, the condition in theorem 17 may be split up into two conditions—absolute integrability in every finite interval $(-R, R)$, which holds for the function of example 24 subject to equation (4) alone; and absolute integrability up to infinity, which in that example requires also (5).

The first condition (absolute integrability in every finite interval) cannot easily be relaxed if we are to have $g(y) \to 0$ as $|y| \to \infty$. Thus, functions satisfying (4) only with $\beta_m \leqslant -1$ do not normally have $g(y) \to 0$, as table 1 shows. The reader should check that all the functions in table 1 with the x^{-m} factor, or the factor $|x|^\alpha$ for $\alpha < -1$, have Fourier transforms which are non-zero or infinite at infinity, while the result about the F.T. of $f(ax + b)$ shows that, if the non-integrable singularity is at $x = -b/a$ instead of $x = 0$, the behaviour of the F.T. as $|y| \to \infty$ is still not a subsidence, but rather a finite or infinite oscillation. Again, the delta function and its derivatives have F.T.'s which are non-zero or infinite at infinity, and so generalised functions which miss being absolutely integrable because of singularities of *this* type will not have their F.T. $\to 0$.

However, table 1 also shows that the second condition (absolute integrability up to infinity, as implied by equation (5) for example) is by no means generally necessary. The reader should check that all the functions in table 1 which are absolutely integrable in every finite interval (namely, those with the x^n factor, or the factor $|x|^\alpha$

for $\alpha > -1$) have Fourier transforms which $\to 0$ as $|y| \to \infty$, even though none of these functions is absolutely integrable up to infinity.

It would be wrong to conclude from this that absolutely any generalised function which is absolutely integrable in every finite interval has its F.T. tending to 0 as $|y| \to \infty$.

EXAMPLE 28. The F.T. of e^{ix^2} is $e^{-i\pi^2 y^2}(1+i)\sqrt{(\tfrac{1}{2}\pi)}$, which does not $\to 0$ as $|y| \to \infty$.

PROOF. The F.T. of the ordinary function $e^{(i-\epsilon)x^2}$, of which e^{ix^2} is easily seen to be the limit as $\epsilon \to 0$, is

$$\int_{-\infty}^{\infty} e^{(i-\epsilon)x^2 - 2\pi ixy}\,dx = e^{\pi^2 y^2/(i-\epsilon)} \int_{-\infty}^{\infty} e^{(i-\epsilon)\{x - \pi iy/(i-\epsilon)\}^2}\,dx$$

$$= e^{\pi^2 y^2/(i-\epsilon)} \sqrt{\left(\frac{\pi}{\epsilon - i}\right)}, \tag{8}$$

which is in the limit as $\epsilon \to 0$ is the function stated.

However, the function of example 28 is somewhat exceptional in that it oscillates with a frequency which itself increases to infinity as $|x| \to \infty$. Most functions occurring in applications do not do this, and the following definition and theorem represent an attempt to include most of them in a general statement without making the latter too complicated to prove.

DEFINITION 20. *The generalised function $f(x)$ is said to be 'well behaved at infinity' if for some R the function $f(x) - F(x)$ is absolutely integrable in the intervals $(-\infty, -R)$ and (R, ∞), where $F(x)$ is some linear combination of the functions*

$$e^{ikx}|x|^\beta, \quad e^{ikx}|x|^\beta \operatorname{sgn} x, \quad e^{ikx}|x|^\beta \log|x|, \quad e^{ikx}|x|^\beta \log|x| \operatorname{sgn} x,$$
$$\tag{9}$$

for different values of β and k.

NOTE. Obviously no values of $\beta < -1$ need be present in $F(x)$.

THEOREM 18. *If the generalised function $f(x)$ is well behaved at infinity and absolutely integrable in every finite interval, and $g(y)$ is its F.T., then $g(y) \to 0$ as $|y| \to \infty$.*

PROOF. It is convenient to divide up the function $F(x)$ of definition 20 into the terms $F_1(x)$ with $\beta > -1$ and the terms $F_2(x)$ with $\beta = -1$. If $F_2(x) = 0$, the proof of theorem 18 is trivial, for $F_1(x)$ is absolutely integrable in every finite interval (it satisfies equation (4)) and therefore $f(x) - F_1(x)$ is; but the latter is given to be also absolutely integrable in $(-\infty, -R)$ and (R, ∞), and hence finally in $(-\infty, \infty)$. Hence, by theorem 17, if $G_1(y)$ is the F.T. of $F_1(x)$,

$$g(y) - G_1(y) \to 0 \quad \text{as} \quad |y| \to \infty. \tag{10}$$

But $G_1(y) \to 0$ by table 1, and so $g(y) \to 0$ as $|y| \to \infty$.

This method fails when $F_2(x) \neq 0$, since $F_2(x)$ is not absolutely integrable in any interval including the origin. However,

$$F_3(x) = \begin{cases} F_2(x) & (|x| > R), \\ \dfrac{F_2(R)(R+x) + F_2(-R)(R-x)}{2R} & (|x| < R), \end{cases} \tag{11}$$

is absolutely integrable in every finite interval (in fact, it is a continuous function), and hence we can conclude as before that $f(x) - F_1(x) - F_3(x)$ is absolutely integrable in $(-\infty, \infty)$. Therefore, if $G_3(y)$ is the F.T. of $F_3(x)$,

$$g(y) - G_1(y) - G_3(y) \to 0 \quad \text{as} \quad |y| \to \infty. \tag{12}$$

But $G_1(y) \to 0$. Also, if $F_{3k}(x)$ signifies the contribution to $F_3(x)$ of terms carrying the e^{ikx} factor, then $F'_{3k}(x) - ik F_{3k}(x)$ is

$$O(|x|^{-2} \log |x|) \quad \text{as} \quad |x| \to \infty$$

and hence is absolutely integrable from $-\infty$ to ∞ (the derivative $F'_{3k}(x)$ has only simple discontinuities at $x = \pm R$) and so its F.T. $(2\pi i y - ik) G_{3k}(y) \to 0$ as $|y| \to \infty$, whence $G_{3k}(y) \to 0$ and, on adding up these results for the different values of k, $G_3(y) \to 0$. Hence finally, by (12), $g(y) \to 0$ as $|y| \to \infty$.

EXAMPLE 29. If $g(y)$ is the F.T. of $f(x) = |x|^\nu J_\nu(|x|)$, where $J_\nu(x)$ is the Bessel function of the first kind, then $g(y) \to 0$ as $|y| \to \infty$ if $\nu > -\frac{1}{2}$.

PROOF. $f(x)$ is continuous except at $x = 0$, where it is $O(|x|^{2\nu})$, so it satisfies equation (4) if $\nu > -\frac{1}{2}$. Hence it is absolutely integrable in every finite interval. It is not absolutely integrable up to infinity, but, by the asymptotic expansion for J_ν,

$$f(x) = F(x) + O(|x|^{\nu - \frac{1}{2} - N}) \quad \text{as} \quad |x| \to \infty,$$

where

$$F(x) = \frac{|x|^{\nu-\frac{1}{2}}}{\sqrt{(2\pi)}} \sum_{n=0}^{N-1} \frac{(\nu+n-\frac{1}{2})!}{n!(\nu-n-\frac{1}{2})!(2i|x|)^n}$$
$$\times \{(-1)^n e^{i(|x|-\frac{1}{2}\nu\pi-\frac{1}{4}\pi)} + e^{-i(|x|-\frac{1}{2}\nu\pi-\frac{1}{4}\pi)}\} \quad (13)$$

is a linear combination of functions of the type (9). It follows that $f(x)$ is well behaved at infinity, since if $N > \nu + \frac{1}{2}$ then $f(x) - F(x)$ satisfies equation (5) and so is absolutely integrable up to infinity. Hence, by theorem 18, $g(y) \to 0$ as $|y| \to \infty$.

We now check this conclusion by calculating $g(y)$. Note first that, if $\nu > -1$, $\lim_{\epsilon \to 0} \{e^{-\epsilon|x|} f(x)\} = f(x)$, since for any good function $F(x)$ we have

$$\left| \int_{-\infty}^{\infty} (1 - e^{-\epsilon|x|}) |x|^{\nu} J_{\nu}(|x|) F(x) \, dx \right|$$
$$\leqslant \epsilon \int_{-\infty}^{\infty} |x|^{\nu+1} |J_{\nu}(|x|) F(x)| \, dx = O(\epsilon) \quad (14)$$

if $\nu > -1$. Now, the F.T. of $e^{-\epsilon|x|} f(x)$ is[*]

$$\frac{2^{\nu}(\nu-\frac{1}{2})!}{\{1+(\epsilon+2\pi iy)^2\}^{\nu+\frac{1}{2}}\sqrt{\pi}} + \frac{2^{\nu}(\nu-\frac{1}{2})!}{\{1+(\epsilon-2\pi iy)^2\}^{\nu+\frac{1}{2}}\sqrt{\pi}}. \quad (15)$$

Hence $g(y)$, the limit by theorem 15 of (15) as $\epsilon \to 0$, is

$$\frac{2^{\nu+1}(\nu-\frac{1}{2})!}{(1-4\pi^2 y^2)^{\nu+\frac{1}{2}}\sqrt{\pi}} \quad \left(|y| < \frac{1}{2\pi}\right), \quad -\frac{2^{\nu+1}(\nu-\frac{1}{2})! \sin \nu\pi}{(4\pi^2 y^2 - 1)^{\nu+\frac{1}{2}}\sqrt{\pi}} \quad \left(|y| > \frac{1}{2\pi}\right). \quad (16)$$

This checks that $g(y)$ does tend to 0 as $|y| \to \infty$ when $\nu > -\frac{1}{2}$; and the fact that $g(y)$ does not tend to 0 for $-1 < \nu \leqslant -\frac{1}{2}$ reconfirms the importance of the condition that $f(x)$ be absolutely integrable in every finite interval.

4.3. The asymptotic expression for the Fourier transform of a function with a finite number of singularities

DEFINITION 21. *A generalised function $f(x)$ is said to have a finite number of singularities $x = x_1, x_2, \ldots, x_M$ if, in each one of the intervals* $-\infty < x < x_1, x_1 < x < x_2, \ldots, x_{M-1} < x < x_M, x_M < x < \infty,$

[*] This follows from Watson's *Theory of Bessel Functions* (2nd ed. 1944), § 13.2, equation (5). Cambridge University Press.

$f(x)$ is equal to an ordinary function differentiable any number of times at every point of the interval.

EXAMPLE 30. $\delta''(x) + |x^4 - 5x^2 + 4|^{-\frac{3}{2}}$ has the singularities $x = -2, -1, 0, 1$ and 2.

Most ordinary or generalised functions which occur in applications have only a finite number of singularities; for these, the method of the present section is very effective. The principal exceptions are periodic functions, which are treated separately in chapter 5.

THEOREM 19. *If the generalised function $f(x)$ has a finite number of singularities $x = x_1, x_2, \ldots, x_M$, and if (for each m from 1 to M) $f(x) - F_M(x)$ has absolutely integrable Nth derivative in an interval including x_m, where $F_m(x)$ is a linear combination of functions of the type*

$$\left.\begin{array}{ll} |x - x_m|^{\beta}, & |x - x_m|^{\beta} \operatorname{sgn}(x - x_m), \quad |x - x_m|^{\beta} \log|x - x_m|, \\ & |x - x_m|^{\beta} \log|x - x_m| \operatorname{sgn}(x - x_m) \end{array}\right\} \quad (17)$$

and $\delta^{(p)}(x - x_m)$, for different values of β and p, and if $f^{(N)}(x)$ is well behaved at infinity, then $g(y)$, the F.T. of $f(x)$, satisfies

$$g(y) = \sum_{m=1}^{M} G_m(y) + o(|y|^{-N}) \quad as \quad |y| \to \infty, \quad (18)$$

where $G_m(y)$, the F.T. of $F_m(x)$, can be obtained from table 1.

PROOF. Let $f(x) - \sum_{m=1}^{M} F_m(x) = f_R(x)$ have F.T. $g_R(y)$. Then $f_R^{(N)}(x)$ has F.T. $(2\pi i y)^N g_R(y)$, and to prove (18) we must show that this $\to 0$ as $|y| \to \infty$. Now, $f_R^{(N)}(x)$ is absolutely integrable in an interval including x_m but no other singularity, because $f^{(N)}(x) - F_m^{(N)}(x)$ is, and so are $F_1^{(N)}(x), \ldots, F_{m-1}^{(N)}(x), F_{m+1}^{(N)}(x), \ldots, F_M^{(N)}(x)$. This being a correct conclusion for $m = 1$ to M, it follows that $f_R^{(N)}(x)$ is absolutely integrable in every finite interval; also, it is well behaved at infinity, since $f^{(N)}(x)$ is given to be, and each component in each of the $F_m^{(N)}(x)$ obviously is. Hence, by theorem 18, the F.T. $(2\pi i y)^N g_R(y)$ of $f_R^{(N)}(x)$ tends to zero as $|y| \to \infty$, as stated in equation (18).

The result of theorem 19 is most often useful when $f(x)$ is an ordinary function. Note, however, that even in this case the statement (18) of the result, let alone its proof, would be meaningless outside generalised-function theory, since in ordinary Fourier-transform theory the transforms $G_m(y)$ would exist only if all the $F_m(x)$ were composed solely of functions of the type (17) with $-1 < \beta < 0$.

The method of theorem 19 will now be illustrated by a number of examples. After studying these, the reader should practise the method on several of the exercises at the end of the chapter. It is instructive to begin with an example to which we already know the answer.

EXAMPLE 31. Find an asymptotic expression for the F.T. of

$$f(x) = |x|^{\nu} J_{\nu}(|x|).$$

SOLUTION. The behaviour of $f(x)$ near its only singularity $x = 0$ is given by the series

$$f(x) = |x|^{\nu} \sum_{n=0}^{\infty} \frac{(\frac{1}{2}|x|)^{\nu+2n}(-1)^n}{n!\,(\nu+n)!}. \tag{19}$$

Hence, if

$$F_1(x) = \frac{|x|^{2\nu}}{\nu!\,2^{\nu}} \tag{20}$$

is the leading term of this series, $f(x) - F_1(x)$ is $O(|x|^{2\nu+2})$ as $x \to 0$, and its Nth derivative is absolutely integrable in an interval including the origin $x = 0$ if N is the least integer $\geqslant 2\nu + 2$. Also, $f(x)$ and its derivatives are well behaved at infinity (see example 29), and so, by theorem 19 and table 1,

$$g(y) = G_1(y) + o(|y|^{-N}) = \frac{\{2\cos\frac{1}{2}\pi(2\nu+1)\}(2\nu)!}{\nu!\,2^{\nu}(2\pi|y|)^{2\nu+1}} + o(|y|^{-N})$$

$$= -\frac{2^{\nu+1}(\nu-\frac{1}{2})!\sin\nu\pi}{(2\pi|y|)^{2\nu+1}\sqrt{\pi}} + o(|y|^{-N}), \tag{21}$$

where $g(y)$ and $G_1(y)$ are the F.T.'s of $f(x)$ and $F_1(x)$, and the 'duplication formula'

$$(2\nu)! = \nu!\,(\nu-\tfrac{1}{2})!\,2^{2\nu}\pi^{-\frac{1}{2}} \tag{22}$$

has been used to throw the result into a form which can be immediately checked from the exact form (16) of $g(y)$.

Note that, by including more terms of the series (19) in $F_1(x)$, one could make higher derivatives (the $(N+2)$th, the $(N+4)$th, and so on) of $f(x) - F_1(x)$ absolutely integrable in an interval including $x = 0$, and so reduce the error in the equation $g(y) = G_1(y)$

for large $|y|$ successively to $o(|y|^{-N-2})$, $o(|y|^{-N-4})$, and so on, depending on how many terms were included. In this way one would build up an 'asymptotic expansion' of $g(y)$, which in the particular case here discussed would be simply the binomial expansion of

$$-\frac{2^{\nu+1}(\nu-\tfrac{1}{2})!\sin\nu\pi}{(4\pi^2 y^2 - 1)^{\nu+\tfrac{1}{2}}\sqrt{\pi}}$$

in descending powers of y, as the reader may check.

EXAMPLE 32. Find an asymptotic expression for the F.T. of $f(x) = |x| \, |x+1|^{\tfrac{1}{2}} \, |x-1|^{-\tfrac{1}{2}}$ with an error $o(|y|^{-2})$.

SOLUTION. The singularities of $f(x)$ are $x = -1, 0, +1$. Expressing $f(x)$ near each singularity as a sum of terms (17), with an error whose second derivative is absolutely integrable in an interval including the singularity, we have

$$\left.\begin{aligned}
f(x) &= \frac{|x+1|^{\tfrac{1}{2}}}{\sqrt{2}} + O(|x+1|^{\tfrac{3}{2}}), \quad f(x) = |x| + O(|x|^2), \\
f(x) &= \frac{\sqrt{2}}{|x-1|^{\tfrac{1}{2}}} + \frac{5|x-1|^{\tfrac{1}{2}}}{2\sqrt{2}}\operatorname{sgn}(x-1) + O(|x-1|^{\tfrac{3}{2}}),
\end{aligned}\right\} \quad (23)$$

as $x \to -1$, 0 and 1 respectively. Taking the right-hand sides of (23) (without the error terms) as $F_1(x)$, $F_2(x)$ and $F_3(x)$ respectively, we see that the conditions of theorem 19 with $N=2$ are satisfied, and therefore (in the notation of that theorem)

$$g(y) = G_1(y) + G_2(y) + G_3(y) + o(|y|^{-2})$$

$$= e^{2\pi i y}\frac{1}{\sqrt{2}}\frac{-\sqrt{(\tfrac{1}{2}\pi)}}{(2\pi|y|)^{\tfrac{3}{2}}} + \frac{2}{(2\pi i y)^2}$$

$$+ e^{-2\pi i y}\left\{\sqrt{2}\,\frac{\sqrt{(2\pi)}}{(2\pi|y|)^{\tfrac{1}{2}}} + \frac{5}{2\sqrt{2}}\frac{(-i\operatorname{sgn}y)\sqrt{(\tfrac{1}{2}\pi)}}{(2\pi|y|)^{\tfrac{3}{2}}}\right\} + o(|y|^{-2})$$

$$= (\sqrt{2})e^{-2\pi i y}|y|^{-\tfrac{1}{2}} - \left(\frac{1}{4\pi\sqrt{2}}e^{2\pi i y} + \frac{5i\operatorname{sgn}y}{8\pi\sqrt{2}}e^{-2\pi i y}\right)|y|^{-\tfrac{3}{2}}$$

$$- \frac{1}{2\pi^2}y^{-2} + o(|y|^{-2}). \quad (24)$$

As in example 31, if higher terms in the expansions of $f(x)$ near each of its singularities were retained in the $F_m(x)$, then the error term in the expression for $g(y)$ could be reduced in magnitude for large $|y|$. One would build up in this way $g(y)$ for large $|y|$ as the sum of three asymptotic series, one in simple powers of $|y|$, one in powers with the factor $e^{2\pi i y}$ outside and one in powers with the factor $e^{-2\pi i y}$ outside.

The 'leading term' in $g(y)$, that is, the one asymptotically biggest for large $|y|$, is that in $|y|^{-\frac{1}{2}}$, which arises from the term in $f(x)$ which is of order $|x-1|^{-\frac{1}{2}}$ as $x \to 1$. This illustrates a general principle, obviously implied by theorem 19 and table 1, that the 'worst' singularity of a function always contributes the leading term to the asymptotic expression for its Fourier transform. (Here, 'worst' is used in the sense that the singularity $x = x_m$, where $f(x)$ is of order $|x - x_m|^{\beta}$, is 'worst' if β is algebraically least.)

Note that the precise order of magnitude of the error term in (24) (instead of the rather vague information that y^2 times it tends to zero) can be deduced from the orders of the error terms in (23); the worst of these is of the $O(|x - x_m|^{\frac{3}{2}})$ form, so its F.T. is $O(|y|^{-\frac{5}{2}})$. We can increase the precision of equation (24), therefore, by replacing the $o(|y|^{-2})$ by $O(|y|^{-\frac{5}{2}})$.

EXAMPLE 33. Find an asymptotic expression for

$$g(y) = \int_0^1 \frac{e^{-2\pi i x y} \cosh x}{(1 - x^4)^{\frac{1}{2}}} \, dx, \tag{25}$$

with an error $o(|y|^{-2})$.

SOLUTION. This $g(y)$ is the F.T. of

$$f(x) = (1 - x^4)^{-\frac{1}{2}} (\cosh x) H(x) H(1 - x), \tag{26}$$

which has the singularities $x = 0$ and 1. Near $x = 0$, $f(x) - F_1(x)$ has absolutely integrable second derivative, being in fact $O(|x|^2)$, if $F_1(x) = H(x)$. Near $x = 1$,

$$f(x) = \frac{\cosh 1 + (x-1)\sinh 1}{(1-x)^{\frac{1}{2}}\{2 - \frac{3}{2}(1-x)\}} H(1 - x) + O(1-x)^{\frac{3}{2}}$$

$$= F_2(x) + O(|x - 1|^{\frac{3}{2}}), \tag{27}$$

where

$$F_2(x) = (\tfrac{1}{2}\cosh 1)(1-x)^{-\frac{1}{2}} H(1 - x)$$
$$+ (\tfrac{3}{8}\cosh 1 - \tfrac{1}{2}\sinh 1)(1-x)^{\frac{1}{2}} H(1 - x). \tag{28}$$

Applying theorem 19 with $N=2$, we deduce by table 1 that

$$g(y) = \frac{1}{2\pi i y} + e^{-2\pi i y}\left\{(\tfrac{1}{2}\cosh 1)\frac{e^{\frac{1}{4}\pi i \operatorname{sgn} y}\sqrt{\pi}}{(2\pi|y|)^{\frac{1}{2}}}\right.$$

$$\left. + (\tfrac{3}{8}\cosh 1 - \tfrac{1}{2}\sinh 1)\frac{e^{\frac{3}{4}\pi i \operatorname{sgn} y}(\tfrac{1}{2}\sqrt{\pi})}{(2\pi|y|)^{\frac{3}{2}}}\right\} + o(|y|^{-2}) \quad (29)$$

as $|y| \to \infty$. From the detailed expressions for the errors in $F_1(x)$ at $x=0$ and in $F_2(x)$ at $x=1$, we can say that the precise order of magnitude of the error in (29) is $O(|y|^{-\frac{5}{2}})$.

EXAMPLE 34. If $F(x)$ and all its derivatives exist as ordinary functions for $x \geqslant 0$, and are well behaved at infinity, derive the asymptotic expansion

$$\int_0^\infty F(x)\sin 2\pi xy\,dx \sim \frac{F(0)}{2\pi y} - \frac{F''(0)}{(2\pi y)^3} + \frac{F^{\mathrm{iv}}(0)}{(2\pi y)^5} - \dots \quad (30)$$

SOLUTION. The 'half-range' Fourier sine integral (30), by §1·4 (see especially equation (18)), signifies $\tfrac{1}{2}ig(y)$, where $g(y)$ is the F.T. of

$$f(x) = F(|x|)\operatorname{sgn} x. \quad (31)$$

Now $f(x)$ has only one singularity, $x=0$, near which $f(x) - F_1(x)$ has absolutely integrable $(2p)$th derivative if

$$F_1(x) = \left\{F(0) + \frac{F''(0)}{2!}x^2 + \dots + \frac{F^{(2p-2)}(0)}{(2p-2)!}x^{2p-2}\right\}\operatorname{sgn} x. \quad (32)$$

Hence the conditions of theorem 19 with $N=2p$ are satisfied, whence, using table 1,

$$g(y) = \sum_{n=0}^{p-1}\frac{F^{(2n)}(0)}{(2n)!}\,2\,\frac{(2n)!}{(2\pi i y)^{2n+1}} + o(|y|^{-2p}) \quad (33)$$

as $|y| \to \infty$. The fact that this is true for all p is what is meant by the 'asymptotic expansion' formula (30) for $\tfrac{1}{2}ig(y)$.

Similarly, under the same conditions, we have

$$\int_0^\infty F(x)\cos 2\pi xy\,dx \sim -\frac{F'(0)}{(2\pi y)^2} + \frac{F'''(0)}{(2\pi y)^4} - \frac{F^{\mathrm{v}}(0)}{(2\pi y)^6} + \dots \quad (34)$$

EXAMPLE 35. Find an asymptotic expression for

$$g(y) = \int_0^1 K_0(x)\cos 2\pi xy\,dx \quad (35)$$

with an error $o(|y|^{-1})$, where K_0 is the modified Bessel function of the second kind of order zero.

SOLUTION. This $g(y)$ is the F.T. of

$$f(x) = \tfrac{1}{2}K_0(|x|)H(x+1)H(1-x) \tag{36}$$

which has singularities $x = -1$, 0 and $+1$, where

$$\left.\begin{aligned}
f(x) &= \tfrac{1}{2}K_0(1)H(x+1) + O(|x+1|), \\
f(x) &= \tfrac{1}{2}\{-\log(\tfrac{1}{2}|x|) - \gamma\} + O(|x|^2\log|x|), \\
f(x) &= \tfrac{1}{2}K_0(1)H(1-x) + O(|x-1|),
\end{aligned}\right\} \tag{37}$$

respectively. Hence, by theorem 19 with $N = 1$ and table 1,

$$g(y) = \tfrac{1}{2}K_0(1)\frac{e^{2\pi iy}}{2\pi iy} - \frac{1}{2}\left(-\frac{\operatorname{sgn}y}{2y}\right) + \tfrac{1}{2}K_0(1)\frac{e^{-2\pi iy}}{-2\pi iy} + o(|y|^{-1})$$

$$= K_0(1)\frac{\sin 2\pi y}{2\pi y} + \frac{1}{4|y|} + O(|y|^{-2}), \tag{38}$$

where the precise form of the error term is $O(|y|^{-2})$ because the 'worst' error term in (37) is of the $O(|x-x_m|)$ form.

EXERCISE 11. Find an asymptotic expression for the F.T. of $e^{-|x|}$ with an error $o(|y|^{-3})$, and check it against the exact expression obtained by direct integration.

EXERCISE 12. Derive the asymptotic expansion of

$$\int_{-\infty}^{\infty} \frac{e^{-2\pi ixy}\,dx}{1+|x|^3}. \tag{39}$$

EXERCISE 13. Find an asymptotic expression for the F.T. of $|x^4 - 5x^2 + 4|^{-\frac{1}{2}}\operatorname{sgn}x$ with an error $o(|y|^{-1})$, and state the precise order of magnitude of the error.

EXERCISE 14. Derive the asymptotic expansion of

$$\int_0^{\infty} (1-x)^{-1}\cos 2\pi xy\,dx. \tag{40}$$

EXERCISE 15. Find an asymptotic expression for

$$\int_0^1 \frac{\log x}{(1-x)^{\frac{3}{2}}}\sin 2\pi xy\,dx \tag{41}$$

with an error $o(|y|^{-2})$, and state the precise order of magnitude of the error.

CHAPTER 5

FOURIER SERIES

5.1. Convergence and uniqueness of trigonometrical series as series of generalised functions

THEOREM 20. *The trigonometrical series*

$$\sum_{n=-\infty}^{\infty} c_n e^{in\pi x/l} \tag{1}$$

converges to a generalised function $f(x)$ if (and only if) $c_n = O(|n|^N)$ for some N as $|n| \to \infty$, in which case the F.T. of $f(x)$ is

$$g(y) = \sum_{n=-\infty}^{\infty} c_n \delta(y - n/2l). \tag{2}$$

Further, the sum $f(x)$ of (1) can be zero only if all the c_n are zero.

From this follows the uniqueness theorem, that *two different trigonometrical series* (1) *cannot converge to the same function* (since then their difference would converge to zero but have non-zero coefficients).

PROOF. By theorem 15, the series (1) converges to a generalised function $f(x)$ if and only if the series (2), obtained by taking Fourier transforms term by term, converges to $g(y)$, the F.T. of $f(x)$. Hence it is sufficient to consider only the convergence of the series (2).

First, *assume* that $c_n = O(|n|^N)$ for some N. Then the 'step function' $g_1(y)$, defined (see fig. 4) as

$$g_1(y) = \left\{ \begin{array}{ll} 0 & (0 \leqslant y \leqslant 1/2l), \\ \sum_{s=1}^{r} c_s & \{r/2l \leqslant y \leqslant (r+1)/2l\}, \\ -\sum_{s=0}^{r-1} c_{-s} & \{-r/2l \leqslant y \leqslant -(r-1)/2l\}, \end{array} \right\} (r \geqslant 1) \right\} \tag{3}$$

satisfies the conditions of definition 7, and so can be regarded as a generalised function. Further, we can write

$$g_1(y) = \sum_{n=1}^{\infty} c_n H(y - n/2l) - \sum_{n=1}^{\infty} c_{-n} H(-y - n/2l), \tag{4}$$

because for any good function $G(y)$

$$\int_{-\infty}^{\infty} g_1(y)\,G(y)\,dy = \sum_{n=1}^{\infty} c_n \int_{n/2l}^{\infty} G(y)\,dy - \sum_{n=0}^{\infty} c_{-n} \int_{-\infty}^{-n/2l} G(y)\,dy,$$

(5)

where the right-hand side of (5) converges even when each term is replaced by its modulus because $G(y) = O(|y|^{-M})$ for all M.

Hence, by theorem 15, the series obtained by differentiating (4) term by term, namely $\sum_{-\infty}^{\infty} c_n \delta(y - n/2l)$, converges to the generalised

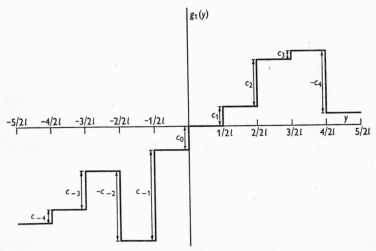

Fig. 4. Graph of the 'step function' $g_1(y)$ of equation (3), whose derivative is $g(y) = \sum_{n=-\infty}^{\infty} c_n \delta(y - n/2l)$.

function $g_1'(y)$, which we may call $g(y)$. Hence (1) converges to the generalised function $f(x)$ whose F.T. is $g(y)$. Note also that, if $f(x)$ were zero, then $g(y)$ must be zero, and so, by theorem 8, $g_1(y)$ is a constant, whence by inspection of (3) all the c_n must be zero.

This simple proof of convergence and uniqueness under the rather unrestrictive condition $c_n = O(|n|^N)$ is all of theorem 20 that is needed for the practical use of trigonometrical series. The proof of necessity of the condition (as stated in the theorem) is so easy, however, that it may as well be given. If there were no N for which $c_n = O(|n|^N)$, then there must be an increasing sequence of n's, say $n_1, n_2, n_3, \dots$, such that $|c_{n_r}| > |n_r|^r$ for each r. Now let

$G(y)$ be any good function such that $G(n/2l) = 0$ if n is not a member of this sequence but $G(n_r/2l) = c_{n_r}^{-1}$. (For example, one might take

$$G(y) = \sum_{r=1}^{\infty} c_{n_r}^{-1} T(2ly - n_r), \qquad (6)$$

where $T(x)$ vanishes for $|x| \geqslant 1$ but is say

$$e^{-x^2/(1-x^2)}$$

for $|x| < 1$ and in particular is 1 when $x = 0$. Then $G(y)$ is a good function by the inequality satisfied by the c_{n_r}.) We may then infer that the series (2) does not converge according to definitions 10 and 11, because

$$\int_{-\infty}^{\infty} G(y) \left\{ \sum_{n=-N}^{N} c_n \delta(y - n/2l) \right\} dy = \sum_{n=-N}^{N} c_n G\left(\frac{n}{2l}\right) \qquad (7)$$

has all its terms zero, except those with n in the sequence $n_1, n_2, \ldots$ which are all 1, and hence it increases without limit as $N \to \infty$. This completes the proof of theorem 20.

DEFINITION 22. *A generalised function of the form* (2) *is called a 'row of deltas' of spacing* $1/2l$. *A generalised function* $f(x)$ *is said to be periodic with period* $2l$ *if* $f(x) = f(x + 2l)$.

EXAMPLE 36. $e^{in\pi x/l}$ is periodic with period $2l$. Hence, by theorem 15, the $f(x)$ of theorem 20 is periodic. Theorem 20 therefore states that the F.T. of a row of deltas of spacing $1/2l$ is a periodic function of period $2l$. (For the converse result, see § 5.3.)

5.2. Determination of the coefficients in a trigonometrical series

We now consider the problem: if we know that

$$f(x) = \sum_{n=-\infty}^{\infty} c_n e^{in\pi x/l} \qquad (8)$$

for some c_n, how can these coefficients be determined? The deeper problem, to prove the existence of such an expansion for any periodic function, is postponed to § 5.3. Note that the classical solution of the present problem (equation (12) of chapter 1),

$$c_m = \frac{1}{2l} \int_{-l}^{l} f(x) e^{-im\pi x/l} dx, \qquad (9)$$

is of no use where generalised functions are concerned, as these cannot be integrated between finite limits.

EXAMPLE 37. No suitable definition of such integration is possible which will give a meaning to $\int_0^1 \delta'(x)\,dx$.

However, the idea of 'integration over a period' can be reproduced by the use of a special kind of function.

THEOREM 21. *A 'unitary function' $U(x)$ can be found, which is a good function vanishing for $|x| \geqslant 1$ and such that*

$$\sum_{n=-\infty}^{\infty} U(x+n) = 1 \tag{10}$$

for all x. The Fourier transform $V(y)$ of any such function has $V(0)=1$, but $V(m)=0$ if m is an integer other than zero.

PROOF. Many such functions $U(x)$ can be found. For any x, at most two terms of the series (10) differ from zero (those with $|x+n|<1$). Therefore, it is necessary only that

$$U(x) + U(x-1) = 1 \quad \text{for} \quad 0 \leqslant x \leqslant 1, \tag{11}$$

and that all derivatives of $U(x)$ should vanish at $x = \pm 1$ (so that they are continuous with zero, their value for $|x|>1$). One may take

$$U(x) = \int_{|x|}^1 \exp\left\{-\frac{1}{t(1-t)}\right\} dt \Big/ \int_0^1 \exp\left\{-\frac{1}{t(1-t)}\right\} dt, \tag{12}$$

for instance (see fig. 5). The exponential ensures that all the derivatives of $U(x)$ vanish (with $U(x)$ itself) at $x = \pm 1$. Condition (11) is easily proved by making the substitution $t = 1 - s$ in the integral. Lastly, if m is any integer,

$$\begin{aligned}
V(m) = \int_{-\infty}^{\infty} e^{-2\pi i m x}\, U(x)\,dx &= \sum_{n=-\infty}^{\infty} \int_{n-\frac{1}{2}}^{n+\frac{1}{2}} e^{-2\pi i m x}\, U(x)\,dx \\
&= \sum_{n=-\infty}^{\infty} \int_{-\frac{1}{2}}^{\frac{1}{2}} e^{-2\pi i m x}\, U(x+n)\,dx \\
&= \int_{-\frac{1}{2}}^{\frac{1}{2}} e^{-2\pi i m x}\,dx = \begin{cases} 1 & (m=0), \\ 0 & (m \neq 0), \end{cases}
\end{aligned} \tag{13}$$

which completes the proof of theorem 21.

The idea of integrating a periodic function $f(x)$ over a period can now be replaced by the idea of integrating $f(x)\, U(x/2l)$ from $-\infty$

to ∞. For in this integral each value of $f(x)$ of the function (which value occurs also at $x + 2nl$ for all integers n) is multiplied by just

$$\sum_{n=-\infty}^{\infty} U(x/2l+n) = 1;$$ but the integration is permissible in the theory

of generalised functions since U is a good function.

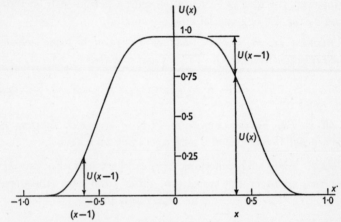

Fig. 5. Graph of the unitary function $U(x)$ of equation (12), illustrating the property $U(x) + U(x-1) = 1$ for $0 \leqslant x \leqslant 1$.

THEOREM 22. *If* $f(x) = \sum_{n=-\infty}^{\infty} c_n e^{in\pi x/l}$, *then*

$$c_m = \frac{1}{2l} \int_{-\infty}^{\infty} f(x)\, U(x/2l)\, e^{-im\pi x/l}\, dx, \tag{14}$$

where $U(x)$ *is any unitary function.*

PROOF. The right-hand side of (14), by definitions 10 and 11, is

$$\lim_{N\to\infty} \sum_{n=-N}^{N} c_n \int_{-\infty}^{\infty} U(x/2l)\, e^{i(n-m)\pi x/l}\, (dx/2l), \tag{15}$$

and the integral in (15) is simply $V(m-n)$, which by theorem 21 is 1 for $n = m$ and 0 for all other n, so that (15) is simply c_m.

5.3. Existence of Fourier-series representation for any periodic generalised function

All the main objects of a theory of Fourier series listed in §1.3 have now been achieved, except that of proving that, if $f(x)$ is any periodic generalised function, and the c_m are defined by (14), then

(8) holds, or (what is the same thing) then $g(y)$, the F.T. of $f(x)$, satisfies $g(y) = \sum\limits_{n=-\infty}^{\infty} c_n \delta(y - n/2l)$. The following theorem is a useful first step towards this, since $\sum\limits_{n=-\infty}^{\infty} U(2ly - n) = 1$.

THEOREM 23. *If $f(x)$ is a periodic generalised function with period $2l$ and F.T. $g(y)$, and if*

$$c_n = \frac{1}{2l} \int_{-\infty}^{\infty} f(x)\, U\!\left(\frac{x}{2l}\right) e^{-\mathrm{i}n\pi x/l}\, dx = \int_{-\infty}^{\infty} g(y)\, V(n - 2ly)\, dy, \quad (16)$$

where the equality of the two forms of c_n follows from theorems 6 and 7, then

$$g(y)\, U(2ly - n) = c_n \delta(y - n/2l). \quad (17)$$

PROOF. We are given that

$$f(x) - f(x + 2l) = 0. \quad (18)$$

Hence, taking Fourier transforms,

$$g(y)(1 - e^{4\pi \mathrm{i}ly}) = 0, \quad (19)$$

which means that

$$\int_{-\infty}^{\infty} g(y)(1 - e^{4\pi \mathrm{i}ly})\, G_1(y)\, dy = 0 \quad (20)$$

for any good function $G_1(y)$. Now, if $G(y)$ is any good function, then

$$G_1(y) = \frac{G(y)\, U(2ly - n) - G(n/2l)\, V(n - 2ly)}{1 - e^{4\pi \mathrm{i}ly}} \quad (21)$$

is also a good function, since the numerator is the difference of two good functions each of which vanishes at all the points $y = m/2l$ where the denominator vanishes, except the point $y = n/2l$, where however each takes the same value by theorem 21. Hence, by (20) and (21),

$$\int_{-\infty}^{\infty} g(y)\, U(2ly - n)\, G(y)\, dy = G\!\left(\frac{n}{2l}\right) \int_{-\infty}^{\infty} g(y)\, V(n - 2ly)\, dy, \quad (22)$$

which proves the theorem.

To complete the proof that $g(y)$ satisfies equation (1) it need only be proved (after equation (17)) that

$$g(y) = \sum_{n=-\infty}^{\infty} g(y)\, U(2ly - n) \quad (23)$$

for any generalised function $g(y)$. This appears almost obvious by theorem 21, but the proof requires some care because our generalised functions are such a very unrestricted class of objects. A lemma on convergence of series is first needed.

THEOREM 24. *If the $a_{m,n}$ are such that $\sum_{n=0}^{\infty} x_n a_{m,n}$ is absolutely convergent and tends to a finite limit as $m \to \infty$ for any sequence x_n which is $O(n)$ as $n \to \infty$, then $\sum_{n=0}^{\infty} \lim_{m \to \infty} a_{m,n}$ converges to the sum $\lim_{m \to \infty} \sum_{n=0}^{\infty} a_{m,n}$.*

PROOF. If the conclusion were false, then an infinite sequence of N's such that

$$\left| \sum_{n=0}^{N} \lim_{m \to \infty} a_{m,n} - \lim_{m \to \infty} \sum_{n=0}^{\infty} a_{m,n} \right| > \epsilon \tag{24}$$

would exist for some $\epsilon > 0$. This means that

$$\lim_{m \to \infty} \sum_{n=N+1}^{\infty} a_{m,n} > \epsilon, \tag{25}$$

or that it is $< -\epsilon$. There must therefore be an infinite subsequence of these N's satisfying only one of these alternatives; the proof proceeds along similar lines whichever it is, but for definiteness suppose it is (25). Let the first of these N's be N_1, and then define the sequences N_s, M_s by induction as follows. If $N_1, ..., N_s$ and $M_1, ..., M_{s-1}$ have been defined (where the former are all members of the sequence satisfying (25)), choose $M_s > M_{s-1}$ (as is possible by (25)) such that

$$\sum_{n=N_s+1}^{\infty} a_{m,n} > \tfrac{1}{2}\epsilon \tag{26}$$

for all $m \geqslant M_s$. Now choose $N_{s+1} > N_s$ as a member of the sequence satisfying (25) such that

$$\sum_{n=N_{s+1}+1}^{\infty} n \, | a_{M_s,n} | < \epsilon. \tag{27}$$

This is possible since $\sum_{n=0}^{\infty} n a_{M_s,n}$ is absolutely convergent.

If now, for each n, x_n is the number of N_s which are less than n, then $x_n = O(n)$ as $n \to \infty$, and so

$$\sum_{n=1}^{\infty} x_n a_{m,n} = \sum_{s=1}^{\infty} \sum_{n=N_s+1}^{\infty} a_{m,n}, \tag{28}$$

because the left-hand series is absolutely convergent. But for $m = M_r$, say,

$$\sum_{s=1}^{\infty} \sum_{n=N_s+1}^{\infty} a_{m,n} > \sum_{s=1}^{r} (\tfrac{1}{2}\epsilon) - \epsilon, \tag{29}$$

by (26) and (27). Hence for the increasing sequence $M_1, M_2, M_3, \ldots$ the series $\sum_{n=1}^{\infty} x_n a_{m,n}$ increases without limit, in contradiction to the hypothesis. This proves the theorem.

THEOREM 25. *If $g(y)$ is any generalised function, and $U(x)$ is a unitary function, then*

$$g(y) = \sum_{n=-\infty}^{\infty} g(y) U(2ly - n). \tag{30}$$

PROOF. By definitions 10 and 11, equation (30) means that, if $G(y)$ is any good function, then

$$\lim_{N \to \infty} \sum_{n=-N}^{N} \int_{-\infty}^{\infty} g(y) U(2ly - n) G(y) \, dy = \int_{-\infty}^{\infty} g(y) G(y) \, dy. \tag{31}$$

To prove this it is necessary to go back to the definition of the generalised function by a regular sequence of good functions $g_m(y)$. Writing

$$a_{m,n} = \int_{-\infty}^{\infty} g_m(y) \{U(2ly - n) + U(2ly + n)\} G(y) \, dy, \tag{32}$$

except for $n = 0$ when the term in curly brackets is taken as only $U(2ly)$ instead of $2U(2ly)$, we can write equation (31) in the form

$$\sum_{n=0}^{\infty} \lim_{m \to \infty} a_{m,n} = \lim_{m \to \infty} \sum_{n=0}^{\infty} a_{m,n}, \tag{33}$$

where on the right-hand side equation (10) has been used.

Now, with the same gloss regarding the term in curly brackets,

$$\phi(y) = \sum_{n=0}^{\infty} x_n \{U(2ly - n) + U(2ly + n)\} \tag{34}$$

is a fairly good function if $x_n = O(n^N)$ as $n \to \infty$ for some N, a result which we use for $N = 1$. (The proof is trivial, since at most two terms of the series can be non-zero for any y.) Hence $\phi(y) G(y)$ is a good function, by theorem 1, and, by the definition of a regular sequence (definition 3),

$$\int_{-\infty}^{\infty} g_m(y)\, \phi(y)\, G(y)\, \mathrm{d}y = \sum_{n=0}^{\infty} x_n a_{m,n} \qquad (35)$$

is absolutely convergent and tends to a finite limit as $m \to \infty$. Hence, by theorem 24, we have equation (33) and hence equation (31) and hence the theorem.

After this digression on convergence matters the main theorem follows at once.

THEOREM 26. *If $f(x)$ is any periodic generalised function with period $2l$ and F.T. $g(y)$, then*

$$f(x) = \sum_{n=-\infty}^{\infty} c_n\, \mathrm{e}^{\mathrm{i}n\pi x/l} \quad and \quad g(y) = \sum_{n=-\infty}^{\infty} c_n\, \delta(y - n/2l), \qquad (36)$$

where
$$c_n = \frac{1}{2l} \int_{-\infty}^{\infty} f(x)\, U\left(\frac{x}{2l}\right) \mathrm{e}^{-\mathrm{i}n\pi x/l}\, \mathrm{d}x. \qquad (37)$$

PROOF. The second of equations (36) follows from theorems 25 and 23, and the first follows from it by theorem 15.

NOTE. If in addition $f(x)$ is absolutely integrable (as assumed in the classical theory of Fourier series) then we have simply

$$c_n = \frac{1}{2l} \int_{-l}^{l} f(x)\, \mathrm{e}^{-\mathrm{i}n\pi x/l}\, \mathrm{d}x, \qquad (38)$$

since for such an $f(x)$ we can write

$$c_n = \frac{1}{2l} \sum_{m=-\infty}^{\infty} \int_{(2m-1)l}^{(2m+1)l} f(x)\, U(x/2l)\, \mathrm{e}^{-\mathrm{i}n\pi x/l}\, \mathrm{d}x$$

$$= \frac{1}{2l} \sum_{m=-\infty}^{\infty} \int_{-l}^{l} f(x)\, U(x/2l + m)\, \mathrm{e}^{-\mathrm{i}n\pi x/l}\, \mathrm{d}x,$$

and then interchange the order of summation and integration. Therefore, the classical Fourier-series theory is included in our more general result.

There is a simple corollary of theorem 26, of little practical importance, but of some general interest.

THEOREM 27. *Under the conditions of theorem 26, c_n is necessarily $O(|n|^N)$ for some N as $|n| \to \infty$. Hence*

$$f(x) = \frac{d^{N+2}f_1(x)}{dx^{N+2}}, \tag{39}$$

$$\left. \begin{array}{l} \text{where} \quad f_1(x) = c_0 \frac{x^{N+2}}{(N+2)!} + \left(\sum_{n=-\infty}^{-1} + \sum_{n=1}^{\infty} \right) c_n \left(\frac{l}{in\pi} \right)^{N+2} e^{in\pi x/l} \end{array} \right\}$$

is a continuous function.

PROOF. By theorem 20, the series (36) which have just been proved to converge could not do so unless $c_n = O(|n|^N)$ for some N. The first of equations (39) is obtained by term-by-term differentiation (permissible by theorem 15). The fact that $f_1(x)$ is continuous follows from the fact that the series for it is absolutely and uniformly convergent, by comparison with the series $\sum_{n=1}^{\infty} n^{-2}$.

The fact that a periodic generalised function must necessarily be a repeated derivative of some continuous function is interesting as showing that there is a limit to the seriousness of the singularities that these functions can have.

We have now derived all the general properties which were noted in § 1.3 as necessary for a satisfactory theory of Fourier series; there exists a unique Fourier-series representation of any periodic function, which converges to the function, whose coefficients can be determined, and which can be differentiated term by term. We conclude with a number of more special results.

5.4. Examples. Poisson's summation formula

This section is mainly concerned with consequences of the following result.

EXAMPLE 38. The generalised function

$$f(x) = \sum_{m=-\infty}^{\infty} \delta(x - 2ml) \tag{40}$$

exists by theorem 20 and is periodic with period $2l$ by theorem 15. Its nth Fourier coefficient is

$$c_n = \frac{1}{2l} \int_{-\infty}^{\infty} \sum_{m=-\infty}^{\infty} \delta(x - 2ml) \, U\left(\frac{x}{2l} \right) e^{-in\pi x/l} \, dx = \frac{1}{2l} \sum_{m=-\infty}^{\infty} U(m) = \frac{1}{2l}, \tag{41}$$

and so by theorem 26

$$f(x) = \frac{1}{2l} \sum_{n=-\infty}^{\infty} e^{in\pi x/l}, \tag{42}$$

an equation whose real form

$$\sum_{m=-\infty}^{\infty} \delta(x - 2ml) = \frac{1}{2l} + \frac{1}{l} \sum_{n=1}^{\infty} \cos \frac{n\pi x}{l} \tag{43}$$

is worth noting; and the F.T. of $f(x)$ is

$$g(y) = \frac{1}{2l} \sum_{n=-\infty}^{\infty} \delta\left(y - \frac{n}{2l}\right). \tag{44}$$

In words, a row of *equal* deltas has as its F.T. a row of equal deltas.

EXAMPLE 39. From (43), by differentiation,

$$\sum_{m=-\infty}^{\infty} \delta'(x - 2ml) = -\frac{\pi}{l^2} \sum_{n=1}^{\infty} n \sin \frac{n\pi x}{l}. \tag{45}$$

NOTE. This is a trigonometrical series on which the old-fashioned 'summability' methods of handling divergent series (which preceded the introduction of generalised functions) broke down. They gave its sum as zero for $x \neq 2ml$ (which is correct, as the generalised function is equal to 0 in any interval not including these points), but also gave it as zero for $x = 2ml$, on the ground that every term of the series vanished at these points. Thus they missed the true character of the singularity at $x = 2ml$; and (what was perhaps even worse) uniqueness was absent in these theories, because of the existence of trigonometrical series whose 'sum' was everywhere zero.

EXAMPLE 40. In the classical theory of Fourier series one calculates the nth Fourier coefficient of the function x in the range $(-l, l)$ as

$$c_n = \frac{1}{2l} \int_{-l}^{l} x e^{-in\pi x/l} dx = \frac{il}{n\pi} (-1)^n, \tag{46}$$

giving

$$\frac{2l}{\pi} \sum_{n=1}^{\infty} \frac{(-1)^{n-1}}{n} \sin \frac{n\pi x}{l} \tag{47}$$

as the full-range Fourier series of x in $(-l, l)$. Now, it is usually stated that such a series cannot be differentiated term by term; and, in fact, the Fourier series of 1 in $(-l, l)$ is not $2 \sum_{n=1}^{\infty} (-1)^{n-1} \cos n\pi x/l$

but $1 + \sum\limits_{n=1}^{\infty} (0) \cos n\pi x/l$. In the theory of generalised functions, however, we have seen that any series can be differentiated term by term. To reconcile the apparent contradiction, note that the sum of (47) is not x, but rather is a periodic function which coincides with x in the period $(-l, l)$. Thus, if

$$f(x) = x \quad (-l < x < l), \qquad x - 2ml \quad \{(2m-1)l < x < (2m+1)l\},$$
(48)

then by theorem 26 and the note following it $f(x)$ is equal to the series (47), and by differentiation

$$f'(x) = 2 \sum_{n=1}^{\infty} (-1)^{n-1} \cos \frac{n\pi x}{l}.$$
(49)

But by (48)

$$f'(x) = \frac{\mathrm{d}}{\mathrm{d}x} \left[x - 2l \sum_{m=1}^{\infty} H\{x - (2m-1)l\} + 2l \sum_{m=-\infty}^{0} H\{(2m-1)l - x\} \right]$$

$$= 1 - 2l \sum_{m=-\infty}^{\infty} \delta\{x - (2m-1)l\}.$$
(50)

This $f'(x)$ is not the function 1. In fact, by equation (43) above, it is

$$-2 \sum_{n=1}^{\infty} (-1)^n \cos \frac{n\pi x}{l},$$
(51)

in perfect agreement with (49).

We learn something useful by applying Parseval's formula to the result of example 38.

THEOREM 28 (Poisson's summation formula). *If $F(x)$ is a good function and $G(y)$ its F.T., then*

$$\sum_{m=-\infty}^{\infty} F(\lambda m) = \frac{1}{\lambda} \sum_{n=-\infty}^{\infty} G\left(\frac{n}{\lambda}\right).$$
(52)

PROOF. This is the equation

$$\int_{-\infty}^{\infty} f(x) F(-x) \, \mathrm{d}x = \int_{-\infty}^{\infty} g(y) G(y) \, \mathrm{d}y$$
(53)

of theorem 6, with f and g as in example 38 (equations (40) and (44)), and $2l$ replaced by λ.

EXAMPLE 41. If $F(x) = e^{-x^2}$, then $G(y) = \pi^{\frac{1}{2}} e^{-\pi^2 y^2}$ and theorem 28 gives

$$\sum_{m=-\infty}^{\infty} e^{-m^2 \lambda^2} = \frac{\sqrt{\pi}}{\lambda} \sum_{n=-\infty}^{\infty} e^{-n^2 \pi^2 / \lambda^2}. \qquad (54)$$

This equation is obvious only for $\lambda = \sqrt{\pi}$. The left-hand side converges very rapidly for $\lambda > \sqrt{\pi}$, and so does the right-hand side for $\lambda < \sqrt{\pi}$—in both cases, faster than

$$1 + 2e^{-\pi} + 2e^{-4\pi} + 2e^{-9\pi} + \dots$$
$$= 1 + 0 \cdot 0864278 + 0 \cdot 0000070 + (10^{-12}) + \dots. \qquad (55)$$

The equation can therefore be used to compute either side very rapidly for any positive λ.

EXAMPLE 42. If $F(x) = e^{-x^2 - 2\pi i x z}$, then $G(y) = \pi^{\frac{1}{2}} e^{-\pi^2 (y+z)^2}$ and theorem 28 gives

$$1 + 2 \sum_{m=1}^{\infty} e^{-m^2 \lambda^2} \cos(2\pi m \lambda z) = \frac{\sqrt{\pi}}{\lambda} e^{-\pi^2 z^2} \left(1 + 2 \sum_{n=1}^{\infty} e^{-n^2 \pi^2 / \lambda^2} \cosh \frac{2\pi^2 n z}{\lambda} \right),$$
$$(56)$$

which is Jacobi's transformation in the theory of theta functions. Similar remarks about convergence apply (the cosh in the series on the right worsens its convergence only slightly, since the expression (56) is a periodic function of z with period λ^{-1} and so one may take $|z| < \frac{1}{2}\lambda^{-1}$ for computational purposes), and the equation is therefore the key to the computation of the theta functions and, through them, of the elliptic functions.

We may note here that Poisson's summation formula (derived in theorem 28 for good functions) can be extended to functions which are not good by approximating to them by a sequence of good functions (as in the proof of consistency of definition 7) and taking the limit. For the formula to be valid in the limit, it is sufficient for $F(x)$ in theorem 28 to be replaced by a function continuous and of bounded variation in $(-\infty, \infty)$ and such that the infinite integral $\int_{-\infty}^{\infty} F(x) \, dx$ converges. In all cases the left-hand side of (52) converges more rapidly for large λ, and the right-hand side for small λ, as in the above examples.

Occasionally the formula leads to a simple analytical form for the sum of a trigonometrical series.

EXAMPLE 43. In the period $0 < z < 2\pi$ we have

$$\sum_{n=1}^{\infty} \frac{\cos nz}{1+n^2} = \frac{\pi \cosh(\pi - z)}{2 \sinh \pi} - \tfrac{1}{2}. \tag{57}$$

PROOF. We may write

$$2 \sum_{n=1}^{\infty} \frac{\cos nz}{1+n^2} + 1 = \sum_{m=-\infty}^{\infty} \frac{e^{imz}}{1+m^2} = \sum_{m=-\infty}^{\infty} F(m), \tag{58}$$

where $\quad F(x) = \dfrac{e^{ixz}}{1+x^2}$ and its F.T. is $\quad G(y) = \pi\, e^{-|2\pi y - z|} \tag{59}$

(as obtained by the method of §3.5). Hence, by Poisson's summation formula,

$$\sum_{m=-\infty}^{\infty} F(m) = \sum_{n=-\infty}^{\infty} G(n)$$

$$= \sum_{n=-\infty}^{0} \pi\, e^{-(z-2\pi n)} + \sum_{n=1}^{\infty} \pi\, e^{-(2\pi n - z)} = \frac{\pi\, e^{-z} + \pi\, e^{z-2\pi}}{1 - e^{-2\pi}}, \tag{60}$$

where the geometric series have been summed for the case $0 < z < 2\pi$. Equation (57) now follows from (58) and (60) by rearrangement.

5.5. Asymptotic behaviour of the coefficients in a Fourier series

The following theorem enables us to use the method of chapter 4 to find the asymptotic behaviour as $|n| \to \infty$ of the Fourier coefficients c_n of a given function $f(x)$.

THEOREM 29. *If $f(x)$ is a periodic generalised function with period $2l$, then $C(y)$, the F.T. of $(2l)^{-1} f(x)\, U(x/2l)$, is a continuous function whose value for $y = n/2l$ is the nth Fourier coefficient c_n of $f(x)$.*

PROOF. By theorem 15 the F.T. of

$$(2l)^{-1} f(x)\, U(x/2l) = (2l)^{-1} \sum_{n=-\infty}^{\infty} c_n\, e^{in\pi x/l}\, U(x/2l) \tag{61}$$

is

$$C(y) = \sum_{n=-\infty}^{\infty} c_n V(2ly - n). \tag{62}$$

This is an absolutely and uniformly convergent series of continuous functions in any finite interval of y, since by theorem 27 $c_n = O(|n|^N)$ for some N but $V(2ly - n) = O(|n|^{-N-2})$ as $|n| \to \infty$.

Hence $C(y)$ is continuous. Also, $C(m/2l) = c_m$ since, by theorem 21, $V(m-n) = 0$ except when $m = n$, and $V(0) = 1$.

NOTE. It is easily shown that $C(y)$ is in fact a fairly good function, but this result is not needed below.

Now, it was noted in connexion with definition 21 that a periodic function $f(x)$ could not have a finite number of singularities (unless the number were zero). However, theorem 29 shows that, provided only $f(x) U(x/2l)$ has a finite number of singularities, then the method of chapter 4 can be applied to determine the asymptotic behaviour of $C(y)$ and hence of the c_n's. Since $U(x/2l)$ vanishes for $|x| > 2l$, the condition is simply that $f(x)$ have a finite number of singularities in any one period.

DEFINITION 23. *The periodic generalised function $f(x)$, with period $2l$, is said to have a finite number of singularities $x = x_1, x_2, ..., x_M$ in the period $-l < x \leqslant l$ if, for some $\epsilon > 0$, $f(x)$ is equal, in each one of the intervals*

$$-l < x < x_1, \quad x_1 < x < x_2, \quad ..., \quad x_{M-1} < x < x_M, \quad x_M < x < l(1+\epsilon),$$

to an ordinary function differentiable any number of times at each point of the interval.

THEOREM 30. *If $f(x)$ is a periodic generalised function, with a finite number of singularities $x = x_1, x_2, ..., x_M$ in the period $-l < x \leqslant l$, such that (for each m from 1 to M) $f(x) - F_m(x)$ has absolutely integrable Nth derivative in an interval including x_m, where $F_m(x)$ is a linear combination of functions of the type*

$$\left. \begin{array}{c} |x-x_m|^\beta, \quad |x-x_m|^\beta \operatorname{sgn}(x-x_m), \quad |x-x_m|^\beta \log|x-x_m|, \\ |x-x_m|^\beta \log|x-x_m| \operatorname{sgn}(x-x_m) \end{array} \right\} \quad (63)$$

and $\delta^{(p)}(x-x_m)$ for different values of β and p, then c_n, the nth Fourier coefficient of $f(x)$, satisfies

$$c_n = \frac{1}{2l} \sum_{m=1}^{M} G_m\left(\frac{n}{2l}\right) + o(|n|^{-N}) \quad as \quad |n| \to \infty, \quad (64)$$

where $G_m(y)$, the F.T. of $F_m(x)$, can be obtained from table 1.

PROOF. The result follows most directly if we take the unitary function $U(x)$ in theorem 29 to be one which equals 1 when $-\frac{1}{2}(1-\epsilon) \leqslant x \leqslant \frac{1}{2}(1+\frac{1}{2}\epsilon)$ and 0 when $x \leqslant -\frac{1}{2}(1-\frac{1}{2}\epsilon)$ or $x \geqslant \frac{1}{2}(1+\epsilon)$. Here, ϵ is that of definition 23, assumed chosen so small that

$x_1 > -l(1-\epsilon)$. In this case, $(2l)^{-1} f(x) U(x/2l)$ is a generalised function with *only* the singularities $x = x_1, x_2, ..., x_M$, and is equal to $(2l)^{-1} f(x)$ in an interval $-l(1-\epsilon) < x < l(1+\frac{1}{2}\epsilon)$ including all of them. It and all its derivatives are 'well behaved at infinity' (they are zero for $|x| \geqslant l(1+\epsilon)$), and so its F.T. $C(y)$, by theorem 19, satisfies

$$C(y) = \frac{1}{2l} \sum_{m=1}^{M} G_m(y) + o(|y|^{-N}) \quad \text{as} \quad |y| \to \infty, \tag{65}$$

whence equation (64) follows by theorem 29.

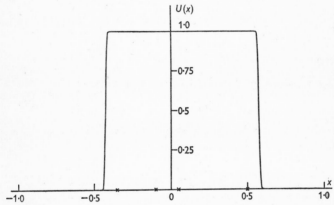

Fig. 6. The unitary function $U(x)$ of equations (66) and (67), in the case $\epsilon = 0.2$. Note that $U(x) + U(x-1) = 1$ for $0 \leqslant x \leqslant 1$, and also that, wherever in the period $-l < x \leqslant l$ the singularities $x_1, x_2, ..., x_m$ are (the crosses indicate possible values of $x_m/2l$), we have $U(x_m/2l) = 1$ for $m = 1$ to M for sufficiently small ϵ.

To complete the proof, we need only produce a unitary function with the stated property. This $U(x)$ (see fig. 6), besides being 0 for $x \leqslant -\frac{1}{2} + \frac{1}{4}\epsilon$ and $x \geqslant \frac{1}{2} + \frac{1}{2}\epsilon$, and 1 for $-\frac{1}{2} + \frac{1}{2}\epsilon \leqslant x \leqslant \frac{1}{2} + \frac{1}{4}\epsilon$, is

$$\int_0^{(2x+1-\frac{1}{2}\epsilon)/(\frac{1}{2}\epsilon)} \exp\left\{-\frac{1}{t(1-t)}\right\} dt \bigg/ \int_0^1 \exp\left\{-\frac{1}{t(1-t)}\right\} dt \tag{66}$$

for $-\frac{1}{2} + \frac{1}{4}\epsilon < x < -\frac{1}{2} + \frac{1}{2}\epsilon$, and

$$\int_{(2x-1-\frac{1}{2}\epsilon)/(\frac{1}{2}\epsilon)}^1 \exp\left\{-\frac{1}{t(1-t)}\right\} dt \bigg/ \int_0^1 \exp\left\{-\frac{1}{t(1-t)}\right\} dt \tag{67}$$

for $\frac{1}{2} + \frac{1}{4}\epsilon < x < \frac{1}{2} + \frac{1}{2}\epsilon$.

Theorem 30, like theorem 19, is most often useful when $f(x)$ is an ordinary function, but again the theorem would be difficult to state without the apparatus of generalised functions. The theorem will now be illustrated by some examples.

EXAMPLE 44. Find an asymptotic expression, with error $o(|n|^{-9})$, for the nth Fourier coefficient c_n of the periodic function

$$f(x) = e^{|\cos x|^3}. \tag{68}$$

SOLUTION. In the period $-\pi < x \leqslant \pi$ the singularities of $f(x)$ are where $\cos x = 0$, that is at $x = \pm \tfrac{1}{2}\pi$. When $x \to \pm \tfrac{1}{2}\pi$,

$$f(x) = 1 + \frac{|\cos x|^3}{1!} + \frac{\cos^6 x}{2!} + \frac{|\cos x|^9}{3!} + \ldots, \tag{69}$$

where

$$|\cos x| = |x \mp \tfrac{1}{2}\pi| - \frac{1}{3!}|x \mp \tfrac{1}{2}\pi|^3 + \frac{1}{5!}|x \mp \tfrac{1}{2}\pi|^5 - \ldots. \tag{70}$$

Hence

$$f(x) = 1 + |x \mp \tfrac{1}{2}\pi|^3 - \tfrac{1}{2}|x \mp \tfrac{1}{2}\pi|^5 + \tfrac{13}{120}|x \mp \tfrac{1}{2}\pi|^7$$
$$+ \tfrac{1}{2}(x \mp \tfrac{1}{2}\pi)^6 - \tfrac{1}{2}(x \mp \tfrac{1}{2}\pi)^8 + O(|x \mp \tfrac{1}{2}\pi|^9) \tag{71}$$

as $x \to \pm \tfrac{1}{2}\pi$. Hence, by theorem 30 with $N = 9$ and $l = \pi$, and table 1,

$$c_n = \frac{1}{2\pi}(e^{-\frac{1}{2}n\pi i} + e^{\frac{1}{2}n\pi i})\left\{\frac{2(3!)}{(in)^4} - \frac{1}{2}\frac{2(5!)}{(in)^6} + \frac{13}{120}\frac{2(7!)}{(in)^8} + o(|n|^{-9})\right\}$$

$$= \frac{12\cos(\tfrac{1}{2}n\pi)}{\pi}\left\{\frac{1}{n^4} + \frac{10}{n^6} + \frac{91}{n^8} + O\left(\frac{1}{n^{10}}\right)\right\}, \tag{72}$$

where the precise form of the error term follows from the detailed expression for the error in (71).

EXAMPLE 45. Find an asymptotic expression for the coefficients b_n in the Fourier sine series for $\sqrt{x}$ in the range $0 < x < l$, with an error $o(n^{-2})$ as $n \to \infty$.

SOLUTION. By §1.3, this Fourier sine series is simply the full-range Fourier series of the odd function $|x|^{\frac{1}{2}}\operatorname{sgn} x$ in $(-l, l)$; and the coefficients b_n in the sine series are $2i$ times the Fourier co-efficients c_n of the periodic function $f(x)$ which equals $|x|^{\frac{1}{2}}\operatorname{sgn} x$ in the period $-l < x \leqslant l$. Now $f(x)$ has singularities at $x = 0$ and l in this period, and

$$f(x) = |x|^{\frac{1}{2}}\operatorname{sgn} x, \quad f(x) = -l^{\frac{1}{2}}\operatorname{sgn}(x - l) + \tfrac{1}{2}l^{-\frac{1}{2}}(x - l) + O(|x - l|^2), \tag{73}$$

as $x \to 0$ (where for once the error in the expression quoted is zero!) and as $x \to l$. Hence, by theorem 30 with $N = 2$, and table 1,

$$b_n = 2ic_n = \frac{i}{l} \left\{ \frac{-i\sqrt{(\frac{1}{2}\pi)}}{(\pi n/l)^{\frac{3}{2}}} - l^{\frac{1}{2}} e^{-n\pi i} \frac{2l}{\pi in} + o(n^{-2}) \right\}$$

$$= \frac{2(-1)^{n-1}\sqrt{l}}{\pi n} + \frac{\sqrt{(\frac{1}{2}l)}}{\pi n^{\frac{3}{2}}} + O(n^{-3}) \tag{74}$$

as $n \to \infty$, where the precise form of the error term follows from the detailed expression for the error in (73).

NOTE. This example exhibits a common feature, where Fourier-series representations of continuous functions in a limited range are concerned—namely, that the 'worst' singularity of that periodic function, which coincides with the given function in the range, is a simple discontinuity at one of the end-points. By theorem 30 and table 1, the nth Fourier coefficient for large n is then necessarily of order n^{-1}. But there are, of course, exceptions to this rule, that is, functions for which no such discontinuity occurs at either end-point (see, for example, exercise 18).

EXERCISE 16. Sum the series $\sum_{n=1}^{\infty} \frac{n \sin nz}{1 + n^4}$ for $0 < z < 2\pi$.

EXERCISE 17. Find an asymptotic expression for the nth Fourier coefficient c_n of the periodic function $|1 + 2\sin x|^{-\frac{1}{3}}$, with error $o(|n|^{-2})$, and state the precise order of magnitude of the error.

EXERCISE 18. Find an asymptotic expression for the coefficients a_n in the Fourier cosine series for $x \log x$ in the range $0 < x < 1$, with an error $o(n^{-3})$, and state the precise order of magnitude of the error.

INDEX